SOUTHERN SURVEYOR

STORIES FROM ONBOARD AUSTRALIA'S OCEAN RESEARCH VESSEL

MICHAEL VEITCH

SOUTHERN SURVEYOR

STORIES FROM ONBOARD AUSTRALIA'S OCEAN RESEARCH VESSEL

MICHAEL VEITCH

PUBLISHING

National Library of Australia Cataloguing-in-Publication entry

Veitch, Michael, author.

Southern Surveyor: stories from onboard Australia's ocean research vessel/
Michael Veitch.

9781486302642 (paperback)
9781486302659 (epdf)
9781486302666 (epub)

Southern Surveyor (Ship) – Travel.
CSIRO – Research.
Marine sciences – Research – Australia.
Scientific expeditions – Australia.

551.46072094

Published by

CSIRO Publishing
36 Gardiner Road, Clayton VIC 3168
Private Bag 10, Clayton South VIC 3169
Australia

Telephone: [+613] 9545 8555
Email: csiropublishing@csiro.au
Website: www.publishing.csiro.au

Front cover: *Southern Surveyor* (MNF/Richard Arculus), waves (MNF/Max McGuire)
Back cover: The Southern Ocean (MNF/Max McGuire)

Edited by Adrienne de Kretser, Righting Writing
Cover design by Andrew Weatherill
Typeset by Desktop Concepts Pty Ltd, Melbourne
Printed by Ingram Lightning Source

Mar26_RP_ILS

Foreword

Australia is in every sense a marine nation. We are surrounded by ocean, over 85% of Australians live within 100 km of the coast, the ocean contributes ~$45 billion per annum (3% of GDP) to our economy, we depend on marine transport for commodity trade, much of our history and cultural ethos is strongly marine-focused, and Australia claims 42% of Antarctica. Australia's marine Exclusive Economic Zone is the third-largest of any nation, and our marine jurisdictional area is 1.8 times larger than our land area (excluding Antarctica). Our climate, and therefore the pattern of rainfall and agricultural productivity, is determined largely by the dynamics of the Southern, Indian and Pacific oceans. Rainfall on Gippsland dairy farms is strongly influenced by the phase of the Indian Ocean Dipole, and whether eastern Australia is flooding or in the grip of drought is influenced by the phase of the El Niño Southern Oscillation in the South Pacific Ocean.

To understand our neighbouring oceans is to understand our geological and cultural history, climate, unique marine biodiversity, economic potential and place in the world. Our future will depend in large part on how we manage human impact on the marine environment, on how we utilise the marine resources within our massive jurisdiction, and on our capacity to predict how human activity will affect the dynamics and functioning of marine systems. This requires a deep knowledge base – it depends fundamentally on an integrative interdisciplinary understanding of marine science across widely varying areas such as physics, chemistry, ocean–atmosphere coupling, marine geoscience, biology, ecology and fisheries science. What a task this is for Australia's marine scientists! And what a crucial role the Research Vessel *Southern Surveyor*, Australia's first truly multi-capable blue-water research vessel, has played in the development of current knowledge.

This strong little ship – at 66 m in length she is only small by international standards – has played a disproportionate role in facilitating Australian offshore marine science. As Australia's only vessel dedicated to blue-water science during her remarkable tenure, and available for use by all Australian marine scientists, her responsibility was large. RV *Southern Surveyor* was built in 1971 as a stern trawler for fishing in the North Sea, but was acquired by CSIRO in 1988 and refitted for fisheries research. In the late 1990s an extensive refit greatly expanded her capability as a multi-purpose research vessel, unique and unprecedented in Australia, after which she was transferred to the Marine National Facility (MNF). Although *Southern Surveyor* was still owned and operated by CSIRO, having the vessel run by the MNF made her accessible to all Australian marine scientists and their international colleagues. The first MNF vessel, RV *Franklin*, had a very specific and narrow capability, so when she was replaced by *Southern Surveyor* there was a massive increase in applications from scientists across government agencies, universities and museums. Competition for access to *Southern Surveyor* skyrocketed. Requests for voyage time to undertake world-class science programs, rated through peer review as 'gold standard', soon far outstripped the 180 voyage days that were funded.

For the next decade until her final voyage in 2013, the *Southern Surveyor* plied the Southern Ocean, the Indian Ocean, waters off northern Australia including Timor-Leste and the Philippines, the Tasman Sea and nearly half-way across the South Pacific Ocean. She facilitated physical, chemical, geological and biological measurements and samples from the sea surface to the ocean depths. She deployed and retrieved state-of-art instruments and tsunami warning buoys, and surveyed vast tracts of seafloor hitherto unmapped. While most voyages were focused on addressing specific questions and hypotheses, others were unashamedly voyages of discovery. And discover they did, turning up new species, new habitats, new canyons, and other geological and physical features previously unknown. The scope and importance of the science she facilitated is breathtaking, and could easily be the subject of a book in itself.

When *Southern Surveyor* completed her last voyage in 2013 it was the end of an era, but at over forty years old she was well past her 'use-by' date. That she provided such grand service for so long is a testament to the quality of the original build and, particularly, to the skill and dedication of the crew. But replacement of *Southern Surveyor* by RV *Investigator* as the new MNF vessel was well overdue and represented a necessary and substantial step-change in science capability and sophistication, and in voyage capability. *Investigator* can carry nearly four times as many scientists for voyages twice as long, and from the tropics to the ice edge. Yet *Investigator*, like *Southern Surveyor*, is funded for only 180 days at sea each year rather than the 300 that would realise her full potential, while demand for access and the need for knowledge would easily occupy three such vessels spending 300 days at sea each year. It is worthwhile noting that Canada, with a similar population size and ocean territory area to Australia, runs sixteen blue-water research vessels over 40 m and up to 98 m in length.

In this eminently readable account, Michael Veitch deftly captures the special nuance, sparkle, excitement and (sometimes) drama of the *Southern Surveyor* story and the extraordinary contribution she has made to Australian marine science. He tells her fascinating story through the eyes of the scientists, crew and scientific and MNF support staff who worked on her. The personalities, the importance of the science, and the affection held for the vessel shine through. It is a fitting tribute that solidifies the special place of *Southern Surveyor* in the history of Australian marine science.

Veitch's entertaining and compelling account also recognises that the need for marine science only escalates as human impacts on marine systems grow, and that there is a need for better knowledge to underpin wise and sustainable use of marine resources and to better understand how oceans affect the climate and earth systems. So it is fitting that in closing the chapter on *Southern Surveyor* he opens the book on *Investigator* as the next step in Australia's powerful contribution to global marine science.

Craig Johnson
Professor, Institute for Marine and Antarctic Science, University of Tasmania
Chair, Marine National Facility Steering Committee 2003–2013
Hobart, April 2015

Contents

Acknowledgements

The journey of this book, much like one of *Southern Surveyor*'s deep sea mooring voyages, has been a long one, and there are several people whose efforts should be spoken of. Foremost, the person whose vision it was to see a tribute to the ship and its people, the unsinkable Sarah Schofield. Her passion in driving the project through the complicated chicanes of CSIRO management has been akin to witnessing a force of nature in itself.

I would like to thank particularly Julia Stuthe and Lauren Webb at CSIRO Publishing whose hushed and gentle tones masked a steely resolve to see the project completed, and whose determination to keep me to my deadlines almost succeeded. Not long into the writing of this book, I sustained a major leg injury, and am grateful beyond words to Jenny Davies for the enormous emotional and physical support she gave me during my recovery, without which I doubt I would have been able to complete the book. Jenny, thank you.

I am indebted to Simon Torok for his fascinating contributions in the boxes and breakout sections, and finally to the many scientists, technicians and ship's crew who gave their time to speak to me so enthusiastically about the *Surveyor* and their work onboard her. Never once did any of you meet my tedious need to have concepts explained, then re-explained, with anything less than politeness and patience. For that, I thank you. This is really your book. You have well deserved it.

Introduction

I only saw her once, at least up close, not long after the curtain had come down after ten years' service as Australia's Marine National Facility. On a cool Hobart afternoon in May, the RV *Southern Surveyor* barely moved on her familiar moorings alongside the CSIRO's Marine Laboratories on an almost still Derwent River. The early winter Tasmanian light had tinged the ship's distinctive deep blue hull with a shade of slate, and her white superstructure seemed worn to a pale grey.

My guide, Mike Jackson, a quietly spoken former naval commander, knew the ship well and was keen to show off every inch of her. As the MNF's Ship Manager, he was also in charge of her disposal.

Beginning at the gangway, Mike led me across *Surveyor*'s once frantically busy back deck. We travelled under her famous stern A-frame, through the laboratories and into the labyrinthine interior, taking in familiar stairwells, flying along now-empty corridors, perhaps for the final time. Something one of her crew had said came back to me. 'She was a small ship, but you could always find more than one way to get from somewhere to somewhere else.'

Walls and linoleum floors were of an indeterminate, neutral colour: a hint of brown, or possibly green. Whether it had worn to this suggestion of a shade, or stubbornly remained the hue chosen for her when put together forty years earlier, was impossible to tell. She was worn, but uniformly so, as if the burden of thousands of hands and feet and the buffeting of countless storms had been smoothed evenly across every surface in a gentle varnish of wear.

Mike was giving me the Grand Tour. Up to the bridge, where stains around ancient window seals showed where water had slowly eaten away at the frames. The rubber viewing slot above one of the radar screens was cracked with tiny capillaries like crow's feet, and had an old smell to it.

We hurtled down corridors, Mike giving reams of information. I put my head inside a room here, a cabin there. Some appeared to have been given a makeover, but long ago, and the imitation wood panelling, bulging oddly in places, had a cosy feel like an old 1960s beach house.

Down we plummeted, deep below *Surveyor*'s water line, into the spectacular noise of the engine room. The generator, simply operating to maintain the ship's electrics while in port, was deafening. 'This is nothing', Mike said. 'You should be in here when the engines are running.' Pipes, cables, valves and steel rods with thick layers of paint created a chaotic Heath Robinson scene, but the calm look on the face of the

RV *Southern Surveyor* off the Great Barrier Reef. Source: MNF.

engineer in heavy-duty earmuffs, who nodded to Mike but eyed me with a hint of suspicion, told me that everything was as it should be.

I was funnelled into the control room, storage areas and galley then under the forecastle, where Mike pointed out the closely constructed iron ribbing of the bow, which helped to give *Southern Surveyor* her phenomenal strength. 'You see', he said quietly, pointing to them, 'just a few inches apart. Immensely strong.' Later, her scientists and crew would attest to that strength, with virtually all of them volunteering that they had never felt unsafe within her sturdy steel walls, even in the severest of storms.

Mike clearly loved the *Surveyor* and this one-off tour was conducted solely to give me both my first and last impressions of her. As we concluded, it struck me that at no time had he referred to the fact that she had already been sold, would soon be gone, and was even now awaiting the arrival of her new owner to take her from her friendly port of Hobart forever. Perhaps he just couldn't bear to think about it.

She was built in 1971, in yard 374 of Brooke Marine Ltd, Lowestoft, England, as the *SN 17* Ranger *Callisto*, the third of four small single-screw stern-trawling fishing factory ships ordered by the now long-vanished Ranger Fishing Co. of North Shields. Like her three sisters, Rangers *Cadmus*, *Calliope* and *Castor*, she was 1106 gross tons, 12 m wide and 66 m in length. Her 16-cylinder English Electric diesel engine was built by Paxman Engineering and delivered 2600 hp to achieve a top speed of 13.5 knots. On her maiden voyage in 1972, she yielded 300 tonnes of freshly caught and quickly frozen fish fillets.

A year later in 1973, during the fishing heyday when the world's oceans were virtually open slather to all comers, Ranger *Callisto*, along with the rest of her Ranger class, was bought by British United Trawlers which, with its tradition of naming its vessels after ancient tribes, called her *Kurd*. She fished until 1981 and experienced at least two dramatic incidents, aiding not one but two British trawlers which had caught fire off the coast of Norway. In 1974 she rescued the crew of the *Victory*, then a year later towed the gutted *Orsino* home from the Norwegian coast.

Hanging over her, however, was the tragic story of her older sister, Ranger *Castor*, renamed *Gaul*, lost at sea in somewhat mysterious circumstances in February 1974. In a severe storm off North Cape Bank, 80 miles north of Norway, all thirty-six crew were lost when she vanished without trace, failing even to issue a distress signal. For those I spoke to who knew the story – and that was by no means everyone – it was as if the *Gaul* subtly haunted *Southern Surveyor*. In quiet tones, the old hands put forward their various theories on her fate.

She was, said some, involved in Cold War espionage. Others that she collided with a Russian or even Western submarine; that she was torpedoed; that she was dragged under by secret undersea cables; that she, according to one engineer, simply capsized due to heavy ice forming on her rigging. Some relatives believe the crew were kidnapped by the Russians and may be alive to this day.

In 1997, the bones of the *Gaul* were finally discovered 270 m down. An investigation concluded that several of her hatches had inexplicably been left open, that she had simply and suddenly been swamped by the storm and went down fast by the stern. For some, though, the mystery will remain.

The introduction of 200 mile limits led to a dramatic shrinking of the British fishing industry in the late 1970s, and *Kurd* was sold again in 1983. The purchaser was a Norwegian company, Vik & Sandvik, with which she would begin a new life as a dive support and survey vessel in the dangerous waters of the North Sea oil fields. *Kurd* underwent a major refit with the addition of a dynamic positioning system (azimuth, bow and stern thrusters), a new and enlarged bridge, a 40 tonne crane and a large moonpool to deploy saturation diving chambers. She was also renamed. *Kurd* became *Southern Surveyor* – an odd choice for a vessel which at that time appeared destined to spend its life exclusively in the waters of the northern hemisphere. It would prove fortuitous, however, when in 1989 she was purchased by Australia's Commonwealth Scientific and Industrial Research Organisation – CSIRO – as its new Division of Fisheries research vessel. The name, naturally, stuck.

Her name, however, would be just about all that would remain, as RV *Southern Surveyor* underwent still more significant conversions, which to some extent were a reversion to her previous life as a trawler. In 1988, upon arrival in Launceston, she was hauled out of the water and her upper observation bridge and large crane were removed, as well as two lifeboats and much of the dive support equipment installed by the Norwegians. Scientific accommodation and laboratories were fitted, and *Southern Surveyor* was back to catching fish. But the times, along with the nature of marine science itself, were starting to change.

The CSIRO Marine Laboratories in Hobart, c. 1988. In the foreground is *Soela*, a stern trawler used by the CSIRO Division of Fisheries. Behind *Soela* is the ocean research vessel *Franklin*, with a blue hull and white stripe. *Southern Surveyor* is pictured with an orange hull, before its refit by the Division of Fisheries. Source: CSIRO.

CSIRO had already put its name to an eclectic fleet of vessels in the name of marine science. There was *Sprightly*, a converted World War II tug and salvage vessel which served for forty years, retiring in 1985. There was *Soela*, a stern trawler used by the Division of Fisheries before her replacement by *Southern Surveyor*.

In 1984, the establishment of Australia's Marine National Facility (MNF) by the Commonwealth led to the construction and commissioning of the 1178 ton ORV (Oceanic Research Vessel) *Franklin*, a purpose-built oceanographic research ship available to all marine researchers working in Australian universities and institutions and to their international collaborators through an independent applications process. Owned and operated by the CSIRO on behalf of the nation, *Franklin* operated widely around Australia as a national facility, largely with oceanographers and geoscientists. The Division of Fisheries vessels, *Soela* and *Southern Surveyor*, focused on fisheries research for CSIRO.

During the 1990s, marine science became more multi-disciplinary, and *Franklin* came to be seen as simply too narrow in scope. Iain Suthers, scientific advisor to the MNF at the time, recalls that after a few voyages on *Franklin*, 'a deep suspicion' set in that she 'really wasn't built for multi-disciplinary work. She didn't have the capability for taking the really heavy loads and doing deep sampling.'

It was also a question of resources, as Tim Moltmann, former Deputy Chief of CSIRO Marine Research, explains. 'The trouble was we had two vessels built for two

RV *SOUTHERN SURVEYOR*: THE NUTS AND BOLTS	
Name	RV *Southern Surveyor*
Port of Registry	Hobart
Research vessel	Multi-purpose
Length overall	66.1 m
Beam	12.3 m
Draft	5.3 m
Gross tonnage	1594
Capacities	Fuel oil 260 tonnes, fresh water 72 tonnes, fresh water evaporator and reverse osmosis up to 10 t/day
Endurance	26 days (5200 nautical miles) at 11 kn
Accommodation	19 single cabins and 5 twin berth
Propulsion	Kort nozzle, controllable pitch propeller, maximum speed 14 kn
Main engine	1 × Wartsila Vasa 6R32E 2460 kW Maximum Continuous Rating at 750 r.p.m.
Thrusters	430 kW bow and stern Brunvoll SPX-VP thrusters, one 552 kW retractable Azimuth unit
Dynamic positioning	2 × Kongsberg Albatross 312
Winches	Main trawl winches 30 t each, net drum winch, coring winch, hydrographic, towed body and scientific winches. The vessel is equipped to deploy demersal and pelagic trawls to 2000 m and to make hydrographic observations to 6000 m
A-frame	6 m above the deck extending over the stern. SWL 15 tonnes and associate general purpose (GP) winch SWL 5 tonnes
Acoustic systems	Kongsberg EM300 multi-beam swath mapper, Simrad TOPAS sub-bottom profiler, Simrad EA500 12 kHz hydrographic sounder, Simrad EK500 38/120 kHz scientific sounder, Sonardyne Ranger USBL system, RDI Ocean Surveyor 75 kHz ADCP
Navigational equipment	Trimble GPS Nav Trac XL and Seapath Seatex 200 GPS systems, Fugro Omnistar DGPS, HSA Endeavour Chart System, 2 × Sperry Mk37 gyro compasses, Simrad AP50 autopilot, 2 × Furuno 96 nm anti-collision radars, Furuno AIS

different purposes, each operating for about half the year.' CSIRO's eyes began to turn towards its existing Fisheries workhorse, *Southern Surveyor*.

In 2002 the rationalisation of the vessels began in earnest with *Southern Surveyor* transferred from CSIRO Marine Research to the MNF to become the new multi-disciplinary vessel. More work was, however, needed for *Southern Surveyor* to keep pace with developments in the science of marine exploration. Her fixed rear trawling gantry was removed and replaced with a vastly more powerful, movable stern

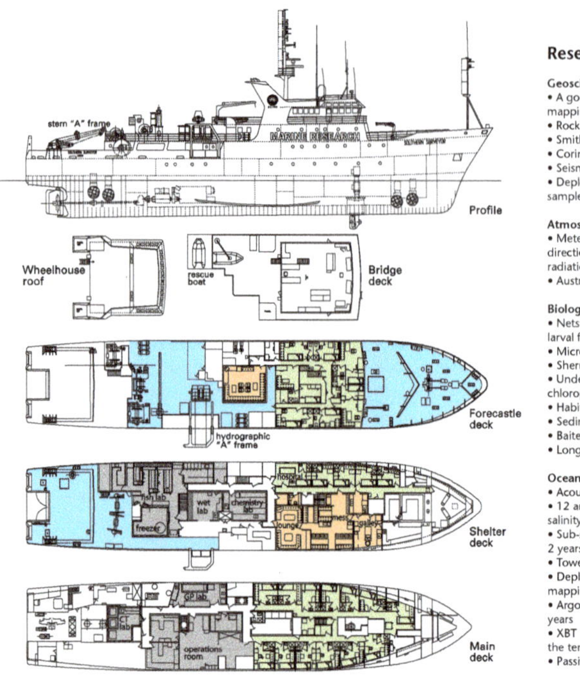

Research capability

Geoscience

- A gondola attached to the hull fitted with seafloor mapping sonar (~3000 metres)
- Rock dredge
- Smith McIntyre Grab
- Coring system
- Seismic surveying
- Deploying remotely operated vehicles (ROVs) for sample and data collection

Atmospheric

- Meteorological instruments – wind speed and direction, air temperature and humidity, rainfall, radiation, atmospheric pressure.
- Australian Tsunami Warning buoys

Biological

- Nets: bongo, surface, drop, mid-water plankton and larval fish, and demersal and pelagic trawl fishing
- Microscopes
- Sherman epibenthic biological sled
- Underway water analysis instruments pCO_2, O_2, chlorophyll and bio-optical sensors
- Habitat mapping with sonar and underwater cameras
- Sediment and fish traps
- Baited underwater cameras
- Long line fishing and tagging for monitoring species

Oceanographic

- Acoustic Doppler Current Profilers (ADCP)
- 12 and 24 bottle CTD water samplers to profile salinity and temperature to monitor ocean currents
- Sub-surface and surface moorings deployed for up to 2 years to collect ocean current data
- Towed ocean profiling sensors – SeaSoar
- Deploying autonomous underwater vehicles (AUV) for mapping and data collection
- Argo robots to collect ocean data for up to eight years
- XBT launch and data acquisition system to measure the temperature profile of the ocean
- Passive sea surface data collection 24 hours a day

Deck plans of the RV *Southern Surveyor.*

A-frame which allowed a wider range of work to be done. Perhaps most significant of all, a large hole was cut into her hull for the installation of a state-of-the-art multi-beam echo sounder, or 'swath mapper'. 'That', said one scientist, was 'like giving a blind man his sight.'

The scientists onboard *Surveyor* now had the ability to see in real time and in three dimensions exactly what was on the seafloor to ~3000 m and later, with the installation of a sub-bottom profiler, even to beneath the seafloor. 'It was absolutely revolutionary', says Tim. 'Before that they were pretty much flying blind or working off charts or what was known about the bathymetry (seafloor terrain), and we could only have done that profiling on the *Surveyor.*'

In 2003, the Divisions of Fisheries and Oceanography merged into the new CSIRO Division of Marine Research. *Franklin* was sold and *Southern Surveyor* stepped into the role of Australia's new multi-purpose MNF vessel.

'It was a tough decision', says Tim, 'and we did go from two vessels to one, which didn't please everybody, but the one we ended up with was far more capable.'

In ten years as Australia's MNF vessel – from 2002 till her final voyage in October 2013 – *Southern Surveyor* would cover 481 550 nautical miles in 111 separate research voyages, her complement of twenty-nine crew, scientists and support staff ranging east into the Pacific Ocean and west to the Indian. She would nudge north of the equator, navigate the torrid currents of the Indonesian throughflow, and sail blithely into the gales of the Southern Ocean to conduct mooring deployments. She would

Each line represents a voyage onboard *Southern Surveyor* during its time as the MNF research vessel (2002–2013). Source: MNF.

carry out oceanographic transects in lines stretching south of New Zealand and north to New Caledonia and Fiji, and work all around the Australian coast.

On the edge of Australia's continental shelf, she would discover vast undersea mountains and ravines that shrink the scale of the Grand Canyon. She would turn up massive underwater volcanoes between Fiji and Samoa and play a role in developing Australia's tsunami early warning system. With her swath mapper, she would detail vast tracts of previously unknown seafloor and prise open the climate records of ancient corals.

She would deploy Integrated Marine Observing System (IMOS) moorings pivotal for collecting data on weather and climate research, calibrate sea surface temperatures given by satellites, observe changes to the ocean currents that influence rainfall and fisheries, and help uncover the southward movement of the East Australian Current. Some of the earliest data she collected would eventually contribute to the global understandings of El Niño and La Niña effects in the Pacific.

She would discover enormous eddies off Rottnest Island and other places, make discoveries that guided mineral exploration worth billions of dollars to Australia, and uncover the final resting place of historically significant shipwrecks. And in what might

In 2006, on a voyage led by Dr Alan Williams from CSIRO, the Tasmanian Seamounts were discovered and mapped in detail. They are located almost due south of Hobart at the base of the continental slope and are an important orange roughy habitat. The seamounts are now part of the Huon Commonwealth Marine Reserve. Source: MNF.

just be a first, she was pivotal in the 'undiscovery' of an island that had remained a phantom on nautical charts for generations.

She was safe and stable, immensely strong, had, say some, 'beautiful lines' and could ride out the worst storms the Southern Ocean could dish up.

But she was old. Her steering and sewerage systems broke down, her prop and Kort nozzle gave trouble, her ageing winches occasionally failed and her hydraulics would leak messily across her back deck. Her engines were replaced, her rusted-out pipes renewed and her electrics upgraded, but as engineer Fred Rostron told me in his dry Lancashire accent, 'She was a bit like grandfather's axe – lasted fifty years but had three new heads and two new handles.'

'In the end', says Tim Moltmann, 'you were a bit worried about doing any work on her because you weren't sure what you were going to discover when you opened something up.' Finally her old bones, like those of all ships, became simply too expensive to maintain.

Southern Surveyor's legacy has been a significant contribution to the sum total of the nation's – indeed the world's – knowledge of our planet, and we are all her beneficiaries. Importantly, in her trials and errors, she also provided the embryo that eventually grew into her stunning successor, RV *Investigator,* Australia's newly built Marine National Facility and the future of Australian marine science exploration.

Toni Moate, CSIRO's Executive Director of the Future Research Vessel Project, puts it simply: '*Southern Surveyor* expanded the concept of how we could deliver multi-disciplinary blue-water science for Australia.'

The purpose of this book is to tell the story of the *Southern Surveyor,* her history and her achievements, not as a narrative but through the voices of the people who knew her and worked onboard her. Over the course of a year, I met physical oceanographers, marine biologists, petrologists and many others who have spent their careers looking at tides and currents, discovering undersea volcanoes or losing sleep over the details of dissolved oxygen levels 3 km below the surface of the Great Southern Ocean. I met the technicians who maintained the scientific equipment, downloaded data, washed out sampling bottles a thousand times and improvised any number of situations in order for the work to continue. I met ship's masters, first mates, voyage managers and engineers, and in the process learned – well, attempted to learn – more about marine science than I could have ever thought possible.

Being someone of a woefully unscientific background, I prepared myself for what I believed would be something of an ordeal by scientific minutiae but instead found myself, time and time again, inspired, even bewitched, by the stories of our oceans, told by passionate people who have dedicated their life's work to learning their secrets, unlocking their mysteries and warning of their perils. I listened as complex ideas were carefully explained: at no time were even my most unscientific questions dismissed or belittled. In fact I never met anything except politeness, courtesy and a genuine desire for me to understand.

I also, of course, learned the story of a ship which, as with all ships, was so much more than an assemblage of steel plates. To many people, *Southern Surveyor* was a laboratory, an office, a home and, above all, a place of discovery.

THE OCEANOGRAPHERS

Tom Trull

Physical oceanographer

'I got off that voyage and just said, "That was horrible".'

Almost everyone I spoke to had, at one stage or another, accompanied physical oceanographer Tom Trull on one of his famous (or perhaps infamous) voyages to the Southern Ocean. They spoke of the experience as one would of having survived some terrible ordeal such as a battle or natural disaster. 'Ah yes …', they would mutter, recalling their saga with Tom and his deep water moorings, turning a little pale with a shake of the head and a slight thousand-yard stare about the eyes. 'Those Tom Trull trips were … something else.' Not that, in terms of the science, Professor Trull's expeditions were ever less than highly successful, yielding important information about the chemical composition of the Southern Ocean. It's just that the weather was sometimes less than perfect.

Tom makes no attempt to sugarcoat the pill. 'My voyages are known as the ones nobody's happy to go on', he says. 'I can't tell you how many students I've seen curled up in a ball for days on end, unable to move.' His first trip, undertaken in 1995, was under the banner 'SPEW', a dubious acronym for 'Southern Ocean Phytoplankton Ecology Watch'. 'But that's not really why we called it that', he confesses. 'We called it SPEW because that's what we spent a great deal of time doing.' Steaming out of Hobart into a vicious storm then plummeting straight down the map to 54° South, the *Surveyor* was 'simply hammered'. 'The *Southern Surveyor* was an absolutely amazing ship in that it could work in really, really rough weather', he assures me, 'but man, could it throw you around!'

On his office door, Tom once pinned up two photographs. The first was an image of the *Surveyor*'s operations room in apparent chaos with chairs strewn to one side and equipment dashed to the floor. The caption read, 'Trull's cruise'. The second presented a contrasting scene of *Surveyor* in a flat calm with a happy, laughing crew steaming through some gorgeous tropical paradise. This caption read, 'Sloyan's cruise', referring to colleague Bernadette Sloyan's expeditions to seas far more benign than the Southern Ocean. The pictures served as a salient reminder to students contemplating a ride with Tom to the south: 'This is what you can expect!'

With the gift of talk seemingly possessed by so many Americans as a birthright, Tom supplies a definitive and stomach-churning description of exactly why, for some, *Southern Surveyor* was a world of motion-sickness pain. 'The ship itself could hold its position in rough weather better than a bigger vessel like the *Aurora Australis*', he explains, 'but it had this motion ...' and he pauses, taking himself back to those hideous hours of being battered by some of the worst seas in the world. 'The bow lifts', he begins slowly, 'and that throws your stomach up towards your mouth. Then it kind of drifts slowly down, and as it does, it sort of corkscrews around. Then just when you think, "Oh it's okay, I'm not going to throw up", it gives a little kick with its tail. And it does that over and over and over again.' I become queasy just listening.

On that first trip, as a junior scientist, Tom found himself tied – quite literally – to his job. His work station was a complicated instrument called a dissolved organic carbon analyser, which worked on a six-minute cycle, at which point the operator was required to perform one of several functions such as loading a sample or observing incoming data. For an entire twelve- to fourteen-hour shift, therefore, every six minutes Tom was beholden to its demanding schedule. 'Barely even time to take a pee', he says. Not that even that brief relief was a realistic proposition, as the weather prevented him from daring to set foot outside the laboratory. 'I was writing software on this little tiny Macintosh computer', he says. 'There was something like a thousand lines of code it was taking from the instrument, and as the ship went up and down I had to scroll up and down looking for tiny errors in the numbers and getting more and more queasy. I had a little blue bucket next to me that I'd throw up into, and I was always near a sink.'

To add to the torture, Tom found the only way to stop himself being thrown around was to take 'a big black and yellow ocky rope' and literally lash himself to the benchtop. 'I got off that voyage and just said, "That was horrible".'

It's somehow refreshing to meet a senior scientist who admits to succumbing to the effects of seasickness. Being a terrible sufferer myself, I marvel at his ability not just to carry on but to think, lead a team and stay focused, when I would simply be curled in the foetal position, willing oblivion. 'You just can't go there', he tells me, 'or else you get stuck there.' The ability to pull yourself out of it, he says, is part physical, part willpower and part brute stubbornness. 'By the beginning of the fourth day I hadn't kept a single meal down and thought, "I've got to keep this inside me or I'm going to be in trouble".' Although every cell in his body fought, that lunchtime Tom went to the canteen, asked for a bowl of soup, forced it down then went to his cabin. 'I lay down on my bunk, clenched my throat, and just waited till it was too far down to come back up again. And from that moment on', he says, 'I was fine.'

Having survived that first trip, eleven months later Tom embarked on his second. He will never forget walking into the laboratory and seeing, still on the floor where he had left it, the little blue bucket he had found such good use for on his first voyage. 'As soon as I saw it', he says, 'my body just said to me, "Throw up!" It was a quite instant, visceral wave of nausea!'

Tom may have gradually got the better of seasickness, but the Southern Ocean weather remained dramatic, sometimes unnerving even the *Surveyor*'s Captain. 'One day', he says, 'we were really struggling. No one was working and the ship was hove-to into a storm, and most people were down in the video room watching a movie and waiting for it to pass.' At the door, the Ship's Master, Ian Taylor, appeared and called Tom into the corridor. 'Look, I just want you to know, I'm just not sure if I can hold the ship.' Tom was slightly aghast: 'I thought to myself, "What on earth does that mean? And anyway, I'm just a scientist!"'

'Well, what do you want me to do?' Tom responded. 'I don't know', said the concerned captain. 'I just thought you'd better know.' Following him up to the bridge, Tom saw that the ship was indeed taking a battering. 'The central wooden frame of the chart table had broken away and they were trying to tape it down', he recalls. 'Was the captain serious?' I ask, 'and what did he actually mean by being unable to "hold" her?' Tom tells me the look on Ian's face said he was indeed quite serious, being concerned that the ship lacked the power to prevent herself being forced beam-on to the waves and 'beaten around' so much that an emergency would ensue. It was perhaps one of the times when the fate of *Surveyor*'s sister ship, *Gaul*, lost with all hands in a North Sea storm, would inevitably come to mind. Despite the drama of the situation, however, Tom says he didn't actually feel unsafe. Eventually, the ship ran back to Hobart for a day before renegotiating the thankfully abating conditions.

From a chemistry undergraduate at the University of Michigan, followed by a PhD via the Woods Hole Oceanographic Institution, then geochemistry at the University of Paris, Tom landed in Hobart in 1993, and for the next twenty years or so sailed with *Southern Surveyor* on 'at least ten voyages, maybe more', working exclusively in the Southern Ocean. I ask what he finds special about this remote and challenging part of the world. His answer is two-fold. 'When I swim in a lake', he explains, 'it's warm on the top and cold on the bottom, and the two layers don't mix very much. It's called stratification. But in the Southern Ocean, that breaks down so you see a lot of interaction between the layers, and between the atmosphere and the deep ocean.'

Another Southern Ocean peculiarity is its inexplicable excess of nutrients and equally inexplicable lack of associated plant life. In most of the world's oceans, there are very low levels of the substances needed to grow plants. 'Like in your garden where you need nitrate and phosphate and so on', he says. In the Southern Ocean, however, this situation is reversed and an abundance of nutrients can be found. Why these are not taken up to create a far greater abundance of tiny plant life has been a subject of ongoing research for twenty years; Tom has concluded that one of the contributing factors is 'limiting levels of iron'.

'Tracking carbon and associated elements and compounds' is how Tom distils his quest for a deeper understanding of the chemical state of the ocean. Aside from moorings and CTDs (extensive hydrographic surveys, generally referred to as CTDs after the primary instrument which measures the water's conductivity, temperature and depth), a considerable armoury of equipment was utilised for the purpose. Some was quite simple, such as sediment traps – basically a big funnel that hung below a surface float – which collected whatever happened to fall from the top of the ocean to the bottom. 'Falling is a way of moving carbon from the surface ocean to the deep ocean', Tom says. 'A little like what ends up at the bottom of your garden at the end of the year. We're interested in all that stuff.' Sometimes it was just a pump on the end of a cable. 'We'd let it pump 10 000 litres an hour through a filter, then bring it up and see what we got.'

Other instruments were more complicated, such as Argo floats – 2 m long, needle-shaped robots thrown into the water, to sink to depths of 2000 m. They zig-zag up and down, measuring velocity, temperature and salinity in the water column before floating to the surface to exchange their data with a satellite, then sink again to repeat the process over and over for a period of eight years. Tom adapted these beyond their typical program of temperature and salinity. 'We deployed floats that had oxygen and bio-optical sensors. One was a blue light to measure chlorophyll – enabling us to tell how many plants are in the water – and the other was a red light to measure the number of particles.'

Gliders were also used, he says, 'like an Argo float with wings'. On a low angle, these glide forward 6–10 km for every 1000 m they travel up and down. 'You throw them in the water and you don't see them again for another six weeks or even up to six months.'

These ingenious devices are actually piloted via satellite technology from a control centre in the University of Western Australia, but sometimes lack the strength to overcome strong currents. 'When we first got them we'd take them out to our site in the Southern Ocean, about thirty-eight hours' steaming south-west of Tasmania, put them in and think they'd just swim home up the Derwent River to Hobart', he says. Fine in theory, but the currents off the continental shelf usually intervened to corral or even send them off course. 'We thought we'd just be able to go out and collect them in a rubber boat; instead we'd have to get a fishing boat and spend a very long day going and getting them.'

Occasionally things went awry, such as when the *Surveyor* ran over its own mooring gear. 'One minute, as we were retrieving it', says Tom, 'we had these nice yellow floats bobbing next to the ship, the next you hear a "klunk" and there's chewed up bits of

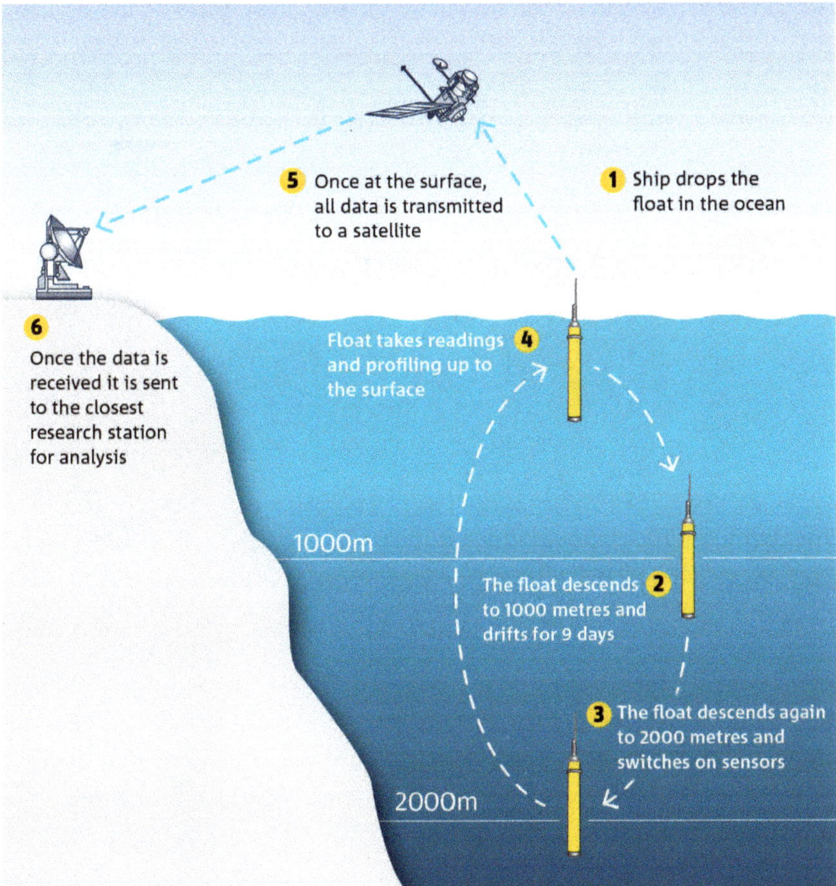

Argo floats are deployed in oceans around the world and can collect and relay data for up to eight years. Source: CSIRO.

plastic going out behind!' At this point, a furious engineer could be heard calling up to the bridge. 'Hey, what the hell are you guys doing up there?!' and the prop, entangled in cable and chain, began to make a disturbing 'wup-wup-wup-wup' sound. 'Well, I don't know if we're going to be able to get it back to Hobart', thundered the engineer as the ship's company became uncharacteristically silent. 'But within an hour', says Tom, 'the thing had cleared itself and we were on our way. We probably lost about $12 000 worth of gear, which is disappointing, but as the whole thing is worth about $300 000 it's not show-stopping. But, yes, it wasn't a brilliant moment!'

It's hard to believe that a man so seemingly attuned to the study of the oceans had for a time given it up altogether. In his first year of graduate school as a junior marine scientist, Tom made a couple of trips to the wild weather of the South Atlantic, where his enthusiasm for the work was somewhat dampened. 'I said, "Stuff this, it's horrible going to sea, I'm not doing that again!", and I switched to volcanology in France. But

On *Southern Surveyor's* final voyage, the scientists and crew retrieve a mooring from the Southern Ocean. Source: MNF/Max McGuire.

the climate problem was getting so important and so exciting, I just thought to myself, "I've got to get back into this".' On that memorable first voyage from Hobart, however, Tom had 'forgotten just how horrible it actually was!' but once onboard the *Surveyor* that was forgotten and he was soon intrigued by some of the burning science questions of the day.

'When I came down there in the early 1990s', he says, 'there was a question as to what caused the fluctuations in CO_2 which registered seasonally in the Southern Ocean.' Was it purely the physics of the warm waters of the East Australian Current charging into the highly mixing cold waters of the sub-Antarctic, expelling their gases 'like a cold beer suddenly warmed inside a stomach?' Tom concluded otherwise. 'We showed the seasonal cycle was actually dominated by biology', he says. 'Although the plants are unable to absorb all the excess nutrients, their growth, death and eventual sinking is enough to be controlling the carbon cycle.'

Once you know it's a biologically controlled process, Tom says, the understanding of climate change becomes a matter not just of physics but also of biology and ecology. 'Then there's the question of this large amount of the dissolved weak acid of CO_2 from fossil fuel emissions beginning to influence the biology. That's become a big part of our work, ocean acidification', he says. 'The ocean is not going to solve the CO_2 problem for us. We thought that it might in the beginning, but now we know that we're going to just have to live with the changes we've made. Now it's a matter of understanding the changes rather than asking, "Will there be any?" That's where scientists are now.'

HOW DO OCEANS SEQUESTER CARBON?

The oceans absorb vast amounts of carbon. They currently remove ~25% of the emissions of carbon dioxide (CO_2) produced by human activities, which is more than 2 billion tonnes of carbon a year. Around 40% of this enters through the Southern Ocean.

This ocean CO_2 'sink' occurs because CO_2 dissolves in ocean waters when atmospheric CO_2 concentrations are higher than those at the ocean's surface. This dissolved carbon stored in the surface waters is mixed down to the deep ocean by the overturning circulation and is locked away from contact with the atmosphere for long periods. Winds, currents and massive whirlpools – known as eddies – that carry warm and cold water around the ocean create localised pathways or funnels that mix the carbon in the deep sea. The waters transported into the deep ocean are replaced by deeper water that has been out of contact with the atmosphere for long periods and can absorb more CO_2. This physical pathway is the main ocean process mopping up our CO_2 emissions and responds directly to increasing CO_2 emissions from human activities because the increase in atmospheric CO_2 results in an increase in the amount of CO_2 the seawater can absorb.

Also, phytoplankton remove carbon dioxide from the surface layer of the ocean when these tiny plants photosynthesise. Just like land plants, phytoplankton convert carbon dioxide and water to carbohydrates and oxygen through photosynthesis. These carbohydrates are then used as fuel by all forms of ocean life. When phytoplankton, and other organisms that have consumed them, die, they sink to the ocean floor, carrying the carbon with them. The carbon can be kept out of circulation for decades to centuries by the dead organisms. While biological pathways have played the dominant role in controlling atmospheric CO_2 levels over millions of years, it is unclear whether in the future they will contribute much to the control of atmospheric CO_2 from human activities.

The carbon sequestration service provided by the oceans comes at a price. A direct result of this CO_2 uptake is the gradual acidification of the oceans. Ocean absorption of CO_2 in the last 250 years has decreased near-surface ocean pH by ~0.1 and is expected to decrease it by a further 0.2–0.3 by 2100. Profound effects on corals and other marine organisms with carbonate skeletons have already been demonstrated, and there is increasing evidence that impacts will occur on a wide range of organisms across the entire marine food web.

Over the years, Tom has noted the evolution of the dynamic between crew and scientists, impressed by the level of engagement showed by the crew in the science work taking place around them, far beyond simply fulfilling the job. 'In the beginning, if I was to be honest', he says, 'it wasn't so much like that. Back then it was like going on a boys' camp.' Over time, however, as crews and scientists spent more time with one another at sea, something akin to a community was slowly forged, the crews becoming more engaged and interested in the work and the scientists gradually respecting the crew's roles and learning the rhythms of a ship.

'When you first get on there, you just think, "Well, I'm just here to do my science"', says Tom. 'Then after a while you start thinking, "Hang on, I can't schedule that particular task for that particular time, it'll run through lunch!" In the last few years, the crews have been amazing, completely capable of working sometimes eighteen-hour days and remaining cheerful about it.'

'When you eat every meal together, pretty intense friendships are forged. And you can't just get up and leave, even if you'd like to. When you think about paths through life', he says philosophically, 'these guys who work as deckies on a ship often didn't get the chances you did of going to university and so forth, and our lives perhaps diverge. But then you're all together on this ship, and all of us love the sea and love learning about it, and you do it together and it's really satisfying.'

Tom reminded me that not all scientists are hands-off when it comes to the physical work required on a voyage. He himself has done everything from firing the pneumatic cannon that shoots the grappling hook out to grab gear, to launching floats and gliders and moving stuff around the back deck. 'We'd get out there and do heavy work, and for guys that spend most of the year in front of a computer, that's pretty rewarding', he says.

On his second voyage, Tom had to climb the main mast to change some filters on a gas intake. 'It was raining and windy and the ship was rocking to the point where you're directly over the water 20 m below, holding on with one hand and turning the fittings on the filter with the other, and forty-eight hours previously, my days had been spent in front of a computer screen! When you're in the lab looking at the data on

An albatross in flight over the Southern Ocean. Source: MNF/Max McGuire.

your computer, you can start to convince yourself that you have a handle on things, but being at sea really reminds you of the scale and the complexity of the unknowns.'

One of his most lingering impressions of being onboard *Surveyor*, however, despite being a chemist, is the birds. 'In my later years', he says, 'there were times I said to myself, "Yeah, I've had enough of this", but I could never get tired of the birds. The albatross, the petrels, and you just look at them and you think, "Wow, we're 500 km from shore and they're just cruising around out here".' He suspects, though, that these stunning animals are a little disappointed with their close encounter with the *Surveyor*. 'They think we're fishermen', he says, 'and they're looking for a feed. They must be thinking, "Gosh, these guys are hopeless, they can't even catch a fish!"'

Bernadette Sloyan

Oceanographer

'If the ocean was opaque I think people would be so much more marvelled by it.'

Oceans within oceans. That's the impression Bernadette Sloyan gives when describing the unbelievably intricate, interconnected systems of currents, temperatures and densities most of us lump blandly together as 'the sea'.

'All the great flows and currents are set up by the way the water interacts with the atmosphere in terms of taking up heat and salt', she tells me. 'It's like a cake with different layers representing different densities and temperatures. You can go down into those layers and read signatures which tell where that water came from and where it was formed.' In the same way one can read atmospheric flows from the pressures on a weather map, an experienced oceanographer such as Bernadette can track the flows of these great water masses over vast distances by determining what temperatures and what degrees of salinity exist within them, and at what depth. How is this remarkable information actually gathered? Well, that's where *Southern Surveyor* comes into the picture.

Although describing herself as a newbie, having been with CSIRO 'merely' a decade, Bernadette has lost count of the number of voyages she's undertaken on *Surveyor* – 'somewhere in the tens or twenties' she says. Many of these were short trips, while others lasted six to eight weeks. 'For six weeks at sea, we collect data that lasts us years', she says. Feeling slightly nauseous at the mere thought of such an extended time at sea, I ask her how she copes. She just throws that one off with a laugh. 'Oh, you get used to it! And hey, for six whole weeks, *you don't have to cook!*'

Bernadette's longer voyages involved extensive hydrographic surveys, or CTDs, as they're generally referred to, after the primary instrument that measures the water's conductivity, temperature and depth. The CTD allows for a detailed examination of a small portion of the ocean. 'I worked mainly in the Pacific', says Bernadette, 'in the basin off New Zealand where it's 6000 m deep.' The CTD probe is a large steel rosette to which an array of bottles and instruments is attached before being lowered to, then raised from, the deep ocean floor. On a typical voyage, up to 100 such CTD casts – each taking up to four hours – are conducted every thirty nautical miles along an imaginary line in the ocean. 'We get a really good look at a slice through the ocean', she says.

I managed to see a CTD just once when *Surveyor* was in harbour. To me it resembled a strange hybrid of high-tech and Heath Robinson – a heavy steel circular frame around which are attached an array of small plastic bottles on a carousel, which are cocked open and shut at certain depths. Its lifeline to the *Surveyor* is a strong steel cable housing a conducting wire. Inside the rosette is the CTD device itself; on more recent voyages, an oxygen probe has been added to the array. As the ship's winch carefully spools out the massive length of cable required to lower it to the deep ocean, the data travels up the conducting cable and into the ship's computer laboratory, where Bernadette and her team of about half a dozen scientists sit watching this live feed of information as it arrives. 'As we watch the CTD go down', she says, 'we see the temperature and salinity profiles change as well as where the water masses are.' Then, on the long ascent, the bottles come into play, fired off in sequence at prescribed depths to collect water samples that are analysed to determine salinity, oxygen and nutrient values, chlorofluorocarbons (CFCs), carbon and other properties that cannot otherwise be measured.

A single twelve-hour shift can yield up to four CTD casts, each one requiring the rosette to be prepared, with bottles and equipment checked and primed, before it is lowered off *Surveyor*'s open side deck. Upon retrieval, the rosette is brought into the *Surveyor*'s wet laboratory where specialised hydrochemists will take water samples for other laboratories to analyse. When satisfied the samples have been successfully taken, the bottles are emptied of remaining water, seals cocked open and made ready for the next descent. As the CTD watcher, Bernadette liaises with the ship's master to signal when the rosette is again ready to be deployed. I visited the *Surveyor*'s wet laboratory only once, and even in dock I got the strong sense it wouldn't be winning too many awards for comfort any time soon. Cramped, noisy, cold and, well, wet, particularly in any kind of sea, it seemed a challenging place to work, like a permanently overflowing laundry attached to the back of a train.

Working on a permanent split twenty-four hour cycle, Bernadette preferred the 2am shift. 'I'd get up at 1am, have five cups of tea and be ready to take over from the people from the shift before me, see where we were, see if there'd been any incidents.' Not only does the data provided by the instruments need to be monitored, but also, vitally, so do the instruments themselves. At several kilometres down, the enormous pressures at which the CTD's specially constructed housings are expected to work render them liable to cracks and damage. 'The CTD has dual temperature and salinity pressure recorders to back each other up', says Bernadette. 'You're constantly watching the integrity of the instruments, and if you think you've got a problem, you

On a voyage to the Indonesian Throughflow with Bernadette Sloyan, students Isabella Rosso (left) and Kate Snow (right) set up the CTD. Source: MNF/Alicia Navidad.

have to switch it out straight away. We may only get to go to some parts of the ocean once every ten years, perhaps only once in our observational lifetime, so maintaining the standard and quality of that data is vital.'

Gradually, painstakingly, every thirty nautical miles, over long and repetitive twelve-hour shifts spent inside the steel walls of a small laboratory at sea for up to six weeks at a time, a detailed picture of the ocean beneath is built up. 'That's when it starts to get exciting', says Bernadette. 'In the long voyages on *Southern Surveyor* we would start at a line at below 50° South and go all the way to the equator, so you see the whole picture change through the entire ocean.'

Some parts of that ocean remain particularly memorable. 'The one I loved the most was the 170° West line', says Bernadette. 'It comes up from Antarctica west of New Zealand and through what we call the Samoan Passage.' South of this passage is a 1000 km wide, 5 km deep ocean basin into which the waters of the Southern Ocean can enter easily, but then are forced into a second basin through a 50 km wide choke point, Bernadette explains.

'Think of it as a big valley in the alps, where all this water is funnelled through at great velocity. It's very exciting. There are waves down there, as big as what we see in the biggest surf, but deep under the ocean.' I mention that I find this somewhat eerie. 'No', she protests with a burst of enthusiasm, 'it's fun! There's so much more mixing and turbulence there for us to look at. The thing about the ocean is that we just look at it from the top, so it seems there is no structure, no hills. If the ocean was opaque I think people would be so much more marvelled by it.'

The CTD is deployed into the Great Barrier Reef during a voyage led by Bronte Tilbrook from CSIRO in 2008, which investigated the vulnerability of ecosystems to acidification. Source: MNF/Robin Beaman.

Bernadette's work on *Surveyor* also involved mooring voyages, trips of shorter duration but no less dramatic in scope. 'We built moorings more than 4 km long anchored to the seafloor and extending to about 20 m below the surface', she tells me. 'Long lines of steel and synthetic rope which we pay out like a big fishing line and attach instruments to measure current, temperature and salinity.'

The *Surveyor*, inching forward at just a couple of knots, must hold the tension on the mooring line until the target zone is reached and the anchor weights are released to plunge kilometres down to their destination. Sea parachutes are fitted to slow the mooring's descent and prevent the impact damaging the line and its attached instruments.

This complex array is left to its own devices to quietly collect data, storing it until it is retrieved and examined eighteen months later. Then, once the instrument casings have been collected, opened and checked, all Bernadette has to do is plug in a lead and start downloading. Within a few hours, she can be reading it all on her screen and that, she says, is a sweet moment indeed. 'You're happy when they go out, and ecstatic when they all come back and you see eighteen months of temperature, ocean velocity and salinity right there in front of you. You always end up saying to yourself, "Wow, there's so much structure there".'

Failure has been rare, but there have been one or two dramatic incidents. 'Yes, we've lost CTDs', she tells me, drawing in a long breath, as if resigning herself to revisiting a painful memory. 'Twice on my voyages in fact, we've lost the whole rosette, although not on *Southern Surveyor.*' I ask her to tell me about it nonetheless. Deploying a CTD in the large swell that constitutes normal conditions in the Southern Ocean, the great steel rosette was being winched from the water to the stern deck when a particularly strong wave emerged from nowhere to give the ship – one of *Surveyor*'s predecessors – a spiteful sideways shove. The winch operator manoeuvred the boom arm as best he could to counteract the sudden slackening of the cable, but it was not enough, and it jumped its sieve-like grate guiding. The ship rose again and the motion snapped the cable like a piece of cotton, sending the rosette and all its equipment straight to the bottom, where it remains to this day. 'It's not a very pleasant thing to happen', says Bernadette, 'but it's very rare.'

Bernadette's passion for what she does becomes evident within seconds of talking with her, but weeks into a long and at times monotonous voyage of exacting twelve-hour shifts, how does she, as Chief Scientist, keep motivated a team of scientists with differing levels of age, experience and limits of endurance? She admits this can be one of the most challenging aspects of the job. 'After six weeks people can lose enthusiasm', she says. 'Especially when you've been on a shift for twelve hours. You just have to keep instilling in the students and the technical staff that this is a long-term process, that it's been a two-year journey to get here: the writing of science proposals, the highly competitive rounds, then the planning and the logistics etc. The payoff will come, but it's at the end, so it's important not to slacken off. After all, it's not as if we can get to just come back again and do the thing tomorrow that we stuffed up today.'

Gaining a sense of what the data is telling you is a long-term process. 'When we go out and do something that we did say ten or twenty years ago and compare the new data with the old and actually see changes in the temperature and salinity of the deep ocean, you really say, "Wow, this is amazing"', says Bernadette. Other phenomena, such as the advance of human-made CFCs, have also been observed. 'We can see CFC concentrations 3000 m down in parts of the ocean that were on the surface and in contact with the atmosphere, say, seventy years ago', she says. 'CFCs show up like a dye in the ocean. We can see their concentrations at the surface of the Southern Ocean below 60° South, and watch them follow a trajectory where they're now registering in the mid latitudes of the Pacific and the Indian and the Atlantic.'

I am reminded of how so few of us think of the ocean as something so complex. Enormous and on the move, certainly, but the more Bernadette talks about it, the more my mind begins to visualise something that is an almost living entity. 'People often think of the deep ocean below 2000 m as being very sluggish and not very dynamic', she says with a hint of wonder. 'We see changes that show it's dynamic from the surface to the deep ocean – a dynamic and interesting place to work in.'

It's also the primary way we track and measure the changes occurring in our climate. 'A vast tracker and incubator of the impact of carbon in the atmosphere and temperature change', she says. 'Because the climate isn't just atmosphere, it's ocean, atmosphere, sea ice, all interacting at different levels and at different timescales. And the oceans are where the warming is happening and it's going to hold that warming for decades and decades. It takes a lot of energy to get heat into the ocean but once it's there, it's there for a long time.'

I suggest to Bernadette that one of the most satisfying parts of her job must surely be the notion of discovery, that sense of extracting real empirical evidence of change in our world with her own eyes. She agrees, but is also conscious of the inherent responsibility being charged with such evidence represents. 'Our challenge is to communicate that to the public', she says. 'That's one of the things I really like about going to sea. I measure these changes and I come back and try to communicate what we're actually observing in the ocean so that people can understand that it isn't just something we play on at the surface but has huge impact on our climate. But yes, it's exciting stuff when you go out and measure it. That's why *Southern Surveyor* has been so important. It's provided us with the means to get that information about how the oceans are circulating.'

North of Darwin, the image of *Southern Surveyor* is reflected perfectly in the still tropical waters. Source: MNF/Alicia Navidad.

Bernadette is one of those lucky souls never to have known the scourge of seasickness, and when at sea on *Surveyor* always opted for the small 'science 1' cabin low down near the bow. 'People usually didn't want that one because it was subject to so much motion, but I loved it', she says, but then she pauses. 'Although it was on the end of the aircon line and sometimes it broke down. Then it wasn't so great, but that didn't happen too often!'

All her years at sea have done nothing to dull Bernadette's passion, indeed her wonder, at the complexity of nature, something she has witnessed first-hand. She leaves me with yet another image, again of the deep, in the Indonesian Throughflow, an immensely important ocean current that transfers warmth between the Pacific and Indian oceans. In stark contrast to the blustery Southern Ocean, these voyages to the north were stifling. No wind, no cooling breeze, a flat calm sea under a tropical sun and the back deck, she says, like a furnace. 'We were working in heavy boots and the heat came up through them into your feet. You really could fry an egg on the deck.'

The mooring here is situated on the northern side of Timor-Leste in the 4000 m deep Ombai Passage. The top instrument sits just 20 m below the surface. Such is the strength of the tide, however, that the whole thing is dragged over, sitting a further 400 m lower. 'It's like a big yo-yo going up and down with this immensely strong tidal current', says Bernadette. 'That's just fascinating to me, to think that the ocean has that much power to push this enormous mooring up and down 400 m. Some people might think that's weird, but nature's just amazing and complex and most of us don't get to see it.'

Nathan Bindoff

Physical oceanographer

'A scientific voyage is a bit of a blind date.'

Nathan Bindoff, scientist, gets straight to the point. 'According to the Myers-Briggs personality types', he tells me out of nowhere on a sunny Hobart day as we sit at a popular cafe across the street from his office at the Institute for Marine and Antarctic Studies at the University of Tasmania, 'only 15% of the population are classified as true introverts, but of that 15%, 85% are scientists.' I laugh at this outrageous statistic and, despite having no idea if it's actually true, I vow to add it to my repertoire of interesting facts. 'Of course', he continues, 'that's not to say that most of the population are scientists', a joke delivered so dryly I almost miss it. Then it's his turn to laugh out loud. 'But seriously, it simply means they (scientists) use intuition to break through logical barriers.' I cannot help think how much intuition Nathan used when contributing to the work which, in 2007, earned for him and his colleagues the Nobel Prize certificate which hangs on the wall of his office.

Despite a jovial exterior, Nathan Bindoff is a serious science heavyweight with a daunting list of qualifications and achievements to his name, as well as being a leading contributor to the science behind global climate change. Professor of Physical Oceanography at the University of Tasmania, and CSIRO Marine Research Laboratories, Climate Change and Ocean Processes Program Leader, Director of the Tasmanian Partnership for Advanced Computing and Project Leader of the Antarctic Cooperative Research Centre's Climate Futures Program and Chief Investigator in the ARC Centre of Excellence in Climate System Science … It's an immensely impressive CV. He was also the coordinating lead author for the ocean chapter in the Inter-Governmental Panel on Climate Change Fourth Assessment Report and Fifth Assessment Report, having documented some of the first evidence for changes in the Indian, North Pacific, South Pacific and Southern oceans. Listing his achievements would run for pages.

He's not bad at telling a good story, either, and we talk for some time. He reveals several rather amazing ocean facts, such as the ability to carbon-date water; that the oldest water is only 1000 years old and found in the depths of the north-east Pacific; that the world's great currents begin a global journey in the North Atlantic, pass down into the South Atlantic and the Southern Ocean before distributing themselves into the Indian and Pacific oceans, then eventually find their way back to their northern hemisphere source, a journey which can take up to 400 years.

The aim of Nathan's July 2013 voyage onboard *Southern Surveyor* was to examine the Southern Indian Counter-current, more specifically to better understand the so-called paradox of the Indian Ocean Gyre. 'It's going in the wrong direction', he says cryptically. 'The wind blows this part of the gyre westwards, but the currents on the surface are driving east.' It sounds a simple enough conundrum, but when Nathan tries to explain it in greater depth using his hands, a faulty pen and a napkin while the waitress cleans up around us, he becomes slightly exasperated, muttering, 'You see? This is why scientists don't write books!' In fact his description is wholly adequate, with an almost poetic picture of the battling natural elements of wind and sea – the ocean surface 'skating against the winds'. Almost Tennysonian, but then it was back to the science.

'If I was to characterise the purpose of the voyages', he says, 'it would be to understand the role of the wind and what I call "surface buoyancy forces" – when warmer water is added to the oceans, driving that eastward flow.' His method utilised Argo floats, drifting, floating sentinels that roam the oceans gathering science data. 'We deployed instruments, we followed eddies, we were looking for an anomalous eastwards flow of the surface waters and the circulation below. Using the satellites is very effective but presents us with a serious mobile phone bill!' The Argo float project is an international effort, with over 3000 floats currently in the world's oceans collecting data.

I was surprised to learn the floats, once deployed, are almost always on a one-way trip, it being far more cost-effective to let them run their roughly eight year cycle of data collection then simply abandon them to the deep or, as sometimes happens, let them be washed up on some remote shore. 'It's a funny feeling throwing $20 000 into the water knowing you're not going to get it back', Nathan tells me. But with the cost of *Surveyor* running at nearly $50 000 per day, it's simple economics.

His second voyage in 2013 followed up the first, and a good deal more Argo floats and low-profile current-measuring 'surface drifters' were deployed, giving top-to-bottom ocean measurements. Each gives a 'full velocity profile through the water column', Nathan tells me. 'Then we get another profile, then another, and because they do it so fast we can get thousands of them which allows us to average out the "noise".' Nathan is still publishing the findings from his 2008 voyage. As he tells me, 'Science is slow. You don't wrap it up quickly. One part of the armour is observation, and observational programs, another part is modelling.'

'A scientific voyage is a bit of a blind date', he says, as his memory of the atmosphere onboard *Surveyor* comes back to him. 'You've got fifteen crew, a scientific party of twelve, some from CSIRO, some from the ship, as well as PhD students, none of whom may have met each other. And there you are, miles from anywhere in the middle of the ocean for twenty-five days working twelve-hour shifts starting midnight and midday.'

And it was hard work. The students, says Nathan, 'didn't quite know what hit them, right from the first day at sea.' Suddenly they were launched into a whole new process involving the getting of measurements, the pressure of having to deliver, having to get it down pat but avoiding becoming complacent 'because that's when you start making mistakes'. He muses a little, and I detect the slightest grin as he recalls the panicked and exhausted young students under his wing, no doubt on the steepest learning curve of their lives, at sea and a long way from home. 'They were a little bit shell-shocked', he confesses with a little sympathy. 'Some get it straight away, and some take the entire voyage.'

Having completed voyages on several research vessels over the years, Nathan is in a position to offer a comparison of their relative virtues. '*Aurora Australis*, the Australian Antarctic Division's icebreaker', he tells me, 'had a nice rolling smooth

The *Southern Surveyor* off the coast of New South Wales. Source: MNF.

motion, while *Southern Surveyor* could be a bit more whippy. However', he adds, shaking his head a little, 'if you could survive both the *Aurora* and the *Surveyor*, you're probably all right to go on the *Franklin*. From a comfort perspective, she was the worst ship I've ever been on!'

Eventually thrown out of the cafe, we let the waitress go home and we retire to Nathan's office at IMAS, adorned as it is with maps, charts and of course his framed Nobel Prize. 'I was only a contributor', he is at pains to point out. He reaches for a volume of the published results of his *Southern Surveyor* voyages, and I am astounded by the meticulousness of the data and its presentation. Every journey of every float and every CTD undertaken on the trip, and the results they gathered, are represented in colourful maps, tables and columns. Once I get the hang of interpretation, I begin to form a portrait of the ocean, as gleaned by Nathan and his team of scientists onboard *Southern Surveyor*.

'This represents the entire trip from end to end', he says, poring over what seems to be the page of an atlas of the ocean, thick with lines of currents, dots, contours and many many numbers. 'These are continuous measurements', he tells me. 'Temperature, salinity, density, oxygen. Every one of those dots represents a place where the ship stopped and samples were physically taken.' With some help and cross-referencing, I ask him to confirm that one particular sample was taken at 45° South at 2500 m. Nathan looks suitably impressed. 'This is what *Southern Surveyor* was doing', he says. 'Contributing to the making of these maps.'

I'm then treated to some further fascinating facts about the ocean, particularly around the Antarctic where some of the coldest, and therefore densest, waters are formed at –2°, so heavy that they 'cascade down the continental slopes and fill up the abyss'. Oceans within oceans, I muse again, as I had done with Bernadette Sloyan.

I ask him in broad terms what was learned by this. 'Boy', he says, pausing, uncharacteristically lost for words. 'How would I sum that up?' By way of answer, he explains to me what was *not* known about the oceans until recently, including much of what he's just told me about the cold waters of Antarctica. 'In the 90s', he says, 'we knew about the structure, shape and overturning circulation of the ocean, but only approximately. Then after a global experiment in the mid 1990s, we knew it in a precise way.' That decade also saw the discovery of new sources of cold dense waters directly to the south of Australia. 'Then it dawned on the world's oceanographic community that these oceans were actually changing, and changing fast.' More recently, he says, it's been discovered that some of these bottom waters are freshening up and getting warmer, and that there's a small de-oxygenation of these waters occurring. This is the sort of work to which *Southern Surveyor* contributed.

I look at the surprisingly modest frame on his wall, which really gives little more than his name, and tell him I've always wanted to know what a Nobel Prize looks like. 'Did you notice something?' he asks. I look again, harder. 'It's not for science, it's for peace', he says, before passing me an enormous volume with two images of the Earth taken from space on its cover. *Climate Change 2007 – The Physical Science Basis – Working Group I Contribution to the Fourth Assessment Report of the IPCC*. 'I wrote this', he says, turning to Chapter 5, headed 'Observations: Oceanic Climate Change and Sea Level. Coordinating Lead Authors, Nathaniel L. Bindoff (Australia), Jürgen Willebrand (Germany)'.

'Well, co-wrote it, to be precise, with Jürgen', he corrects. I start to read the chapter's summary, and stop at the first line which states, quite simply, 'The oceans are warming.' I tell him that really is quite an opening, and he chuckles. 'They're enormous efforts by the climate community globally.' Along with 'a bunch of others', Nathan went to Sweden for four days and, with government representatives, thrashed out the all-important chapter summary 'line by line, word by word'. On day three, he tells me, they worked through the night, and on day four through the entire day. 'So some people basically hadn't slept for two days. But if you don't agree on everything, you run out of time, and it all just spills onto the floor.'

'Scientists are privileged. When you think about it, it's a privilege. But I'm an optimist. I really do remain optimistic that we will mitigate climate change.'

Anya Waite

Oceanographer

> *'I built my career as an international oceanographer on the* Southern Surveyor.*'*

I spoke to Anya Waite at her new home in Germany, via the modern miracle of Skype. I'd used Skype a couple of times and thought I'd become pretty used to it, but really as little more than a cheap phone line with pictures, which always seem rather redundant to me anyway. As we spoke, however, someone happened to walk past her window. I could see it was a cloudy day in Bremerhaven and there was just the briefest shadow of a passing shoulder, but for some reason it registered as one of those moments that speak of how the world has changed so very very much, so very very quickly. 'I just saw a person walk past somebody's window … in Germany' rang the sonorous realisation in my head. It made me wonder, for a moment, if I had not perhaps just woken from a thirty-year coma to find people travelling in cars without wheels.

Somehow, I doubt whether a person with a brain as impossibly large as Anya Waite's would be quite so nonplussed by such ho-hum, everyday technology. Besides, some of the things she told me were pretty astounding too.

'In 1997 I came over from Canada to the University of Western Australia to bring a flavour of applied ecology to their engineering school', she tells me, rendering me quite silent. 'Sorry, can you just say that again?' I ask. 'Well, it's not as far-fetched as it may sound', she says, perhaps picking up my general air of bewilderment. 'As an oceanographer you have a similar training to an engineer – you learn a lot of maths and physics – and then you use that as a framework for ecological and chemical processes. Putting natural processes into their physical context is what an oceanographer does, and it's what engineers do, they just use different scales. I was taking my oceanography training and using it in things like urban drainage and wetlands management.'

My reaction to this must have been a long silence, as Anya suddenly burst into a bright and unexpected laugh. 'It's not as complicated as it sounds, I promise!'

More out of curiosity than anything else, I ask Anya if she could briefly outline her academic history. This time the pause comes from her end. 'Do you *really* want me to?', she asks. I listen to a catalogue of achievements. '...English and biology as a double major ... a minor in music ... an honours degree in forestry ...' I lose track of some of the minor accolades but at some point Anya found time to squeeze in an oceanography degree, and it set her on her course. 'As soon as I went to sea', she says, 'I fell in love with the ocean and the colour blue – and that was it.'

From there, Anya proceeded to a PhD in Vancouver, followed by a postdoctorate at the famous Woods Hole Oceanographic Institution from where, becoming disillusioned with the competitive nature of American science ('as a Canadian, I come with a built-in anti-American gene', she warns), Anya moved to New Zealand to work as a specialist in confocal microscopy, building 3D images of plankton.

Sea time with the New Zealand research vessel *Tangaroa* ensued, where Anya took part in 'one of the most important science research cruises of the 20th century', in which a vast artificial plankton bloom was created and tracked by dumping 7 tons of iron sulfate in the traditionally iron-poor waters of the Southern Ocean. 'That really put southern hemisphere ocean science on the map', she says. Her unique skill set was picked up by the University of Western Australia where, she drops airily, 'I had to train myself to be an engineer so I could teach engineers'. As you do.

Heading out to sea 'a few times', as one of the science team on Marine National Facility voyages with the *Franklin*, Anya was struck by what she describes as 'the quality of these really nice Australian ships and the support staff doing heaps of oceanography that no one else was doing'. Interesting, yes, but Anya was developing serious research ideas of her own, and by 2003 it began to dawn on her that if she were ever going to do the sort of voyages she really wanted to do, she was going to have to lead them herself.

'I had this biophysical coupling obsession where I really wanted to look at the scales of how physics and biology interact, and nobody was doing measurements on the scales I wanted them to do. So finally I just said, "Well, I better just write my own ship time proposal" and I actually got it!' she says, still slightly amazed by the fact that the MNF decided in 2003 to grant her a voyage as Chief Scientist aboard the *Southern Surveyor*. 'It was just the most amazing experience', she says. 'Totally transformative for a young female scientist, to get to go to sea and run the show and have such incredible support from the MNF.' Even though she confesses to 'not really knowing what it all entailed' she was soon to find out, as was the rest of the world.

'Whirlpool off Rottnest!' shouts the newspaper *West Australian*'s headline in June 2006. 'A massive, 200 km wide whirlpool had been discovered by scientists west of Rottnest, creating a balmy climate over the sea surface ...'. As it makes its way down the Western Australian coast carrying heat and nutrients from the tropical north, the Leeuwin Current arrives off Perth and starts to form large rotating water masses, or eddies. 'They're like an egg-beater', says Anya. 'They form, they start to rotate, then they kind of get a life of their own.' And they're huge – the largest of them up to

200 km in diameter and reaching down beneath the surface a full kilometre. As the newspaper article attests, it was these eddies that Anya managed to discover in the *Southern Surveyor*: 'Moving at 6 km/h, the giant whirlpool is strong enough to drag a 70 m CSIRO ocean research vessel. It is thought to form each autumn as the south-flowing warm Leeuwin Current gathers strength close to the WA coast.' Until she found them nobody even knew they were there. The article quotes a local Sea Search and Rescue operation manager as saying, 'First I've heard of them.'

The media smelt a story, and a good one. 'It was hilarious', says Anya, chuckling still. 'In one of those funny moments in an interview, I referred to it as "The Vortex". Straightaway it caught the imagination of the press.' The story of possibly sinister vortexes lurking off the Western Australian coast raced around the world, being picked up by the London *Times*, *Scuba Diving* magazine and a host of other outlets fascinated by the image of spiralling whirlpools swallowing ships in their deadly grip. Blockbuster stuff.

Press frenzy aside, these giant eddies are indeed fascinating, forming out of the currents and heading westward under their own power until they are gradually dissipated by friction. Something in this description prompts me to suggest to Anya the 'Tasmanian Devil' character from the Warner Brothers' cartoons – it swirls around in its own private tornado, creating mayhem. She laughs all the way from Bremerhaven.

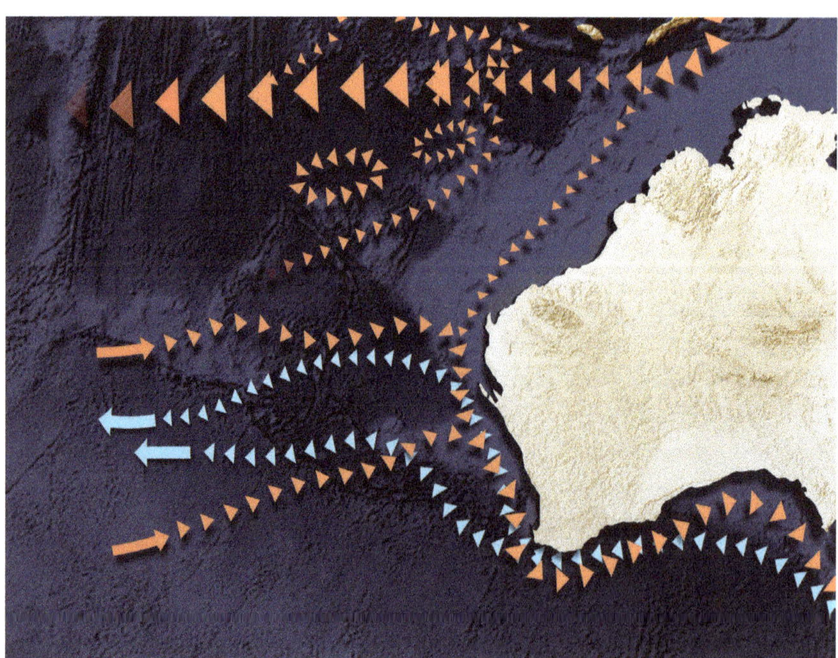

Ocean currents to the west of Australia. Source: CSIRO, the Wealth from Oceans Flagship and the Australian Climate Change Science Program.

The SeaSoar is lowered over the back deck of *Southern Surveyor* using the A-frame. Source: MNF.

Anya used a variety of means in her study, including CTD casts, moorings deployed in the eddy's swirling centre, nets to study the biology and, of course, the famous – now lost – SeaSoar. CSIRO technician Lindsay Pender, who Anya describes as her 'secret weapon', came on the voyage to operate the SeaSoar and, in the name of science, pushed it to its limits. 'At one point', she says, 'the SeaSoar mapped the bottom of the warm-core eddy, which went down to about 400 m.' This was completely outside what the instrument was supposed to be able to do and it didn't seem to like it, shaking, vibrating and clearly not happy. The experiment was too risky to repeat, but after it had been done, Anya says, 'Everyone felt a sense of complete and utter triumph.' I would hear a good deal more about the SeaSoar, including the story of its sorry demise, when I spoke to Lindsay.

The voyage was also remarkable not so much for the loss of gear – which inevitably happened on the *Surveyor*, as it does on all ships from time to time – but for finding it again. One of the large buoys deployed in the centre of the vortex, to track its trajectory, was a new and untested example – $21 000 worth, Anya tells me – and 'Nobody was even sure if it was going to work.' Her hunch seemed correct, as shortly after it was launched and sailed away, it went silent and stopped signalling. They looked but failed to find it, despite it being fitted with a solar-powered strobe. 'It really hurt', says Anya, 'to have lost this buoy – $21 000 to a young investigator to whom every cent is a gift.'

Three days later, the buoy long gone, the *Surveyor* was making its way back through the centre of the eddy at night, when a voice went up from the bridge, 'I see the strobe.' Anya and everyone else onboard were stunned. 'We'd travelled 100 km in the other direction', she says.

This episode illustrated one of the eddy's characteristics. 'It showed us that these eddies are strongly surface convergent', says Anya, 'so there's this one spot right in the middle where everything is going to collect. And there it was.' They scooped it up, got their $21 000 back, fixed its faulty battery and the bloke who spotted it was given a slab of beer of his choosing – on return to port.

The eddies explored by Anya off Western Australia are in fact a global phenomenon, occurring in active boundary currents in places like the Antarctic Circumpolar Current, the Gulf Stream and Africa's Agulhas Current, but it is only in the last ten years or so, according to Anya, that they have been studied in depth. 'We tracked them at sea and tried to find the centre of them using local current information and satellite images, then we mapped them. That had never really been done before in Australia. In fact', she says, 'we were a little bit ahead of our time going into the dynamics of these things way before people had woken up to them.'

Anya participated in eight *Southern Surveyor* voyages, six as Chief Scientist investigating such diverse subjects as reef/ocean interactions at Ningaloo Reef, and two trips exploring the dynamics of rock lobster fisheries off Western Australia. 'Rock lobsters are amazing creatures', she tells me. 'At the larval stage they look like aliens, then they metamorphose into tiny baby lobsters in twenty minutes!' One evening, while watching one of the larvae form what appeared to be stripes on its legs, they were called to dinner, and on their return found that it had transformed from 'a tiny, flat glass spider' into a perfect little cray.

These rock lobster voyages contributed significantly to the overall knowledge of the subject, discovering, for example, that the tiny creatures' dining habits favour a particular type of planktonic worm called chaetognatha or arrow worm, that they have a curious habit of sitting on, while eating, tiny jellies or salps, and that the eddies Anya had learned about earlier were important to their development. 'In certain types of eddies', she says, 'we found they have more lipids and are more likely to be healthier. We were very lucky. We managed to publish eight or nine papers just from that rock lobster work.'

By its very nature, oceanographic work is not easy, particularly, says Anya, for female scientists such as herself. 'I think women oceanographers really struggle. The first time I went to sea in 2003, I had to leave a fifteen-month-old at home. She had a great dad and excellent daycare, but to get this opportunity to run my cruise at this stage in my career, I felt I had to make that sacrifice.' I joke that when Anya lines up to receive her Nobel Prize her daughter may forgive her, at which she laughs, but I was struck by the fact that it was a subject which had not been touched on by any other scientist I had spoken to, male or female.

Anya is acutely aware of the opportunity awarded her by the Marine National Facility, and could not be more fulsome in her praise of the *Surveyor* and her crew. 'Not only did they do it', she says, 'but they really came onboard with it. We had the First Mate jogging up to the current profiler and making sure we were in the right place, and getting into the mapping and tracking. It was just the most amazing experience, totally transformative for a young female scientist, to get to go to sea and run the show and have such incredible support.'

From the three-and-a-half week eddies voyage in October 2003 alone, Anya and her team were able to produce fifteen international journal papers, and in 2007 a special edition of the prestigious *Deep Sea Research* focused on her work on the Leeuwin Current and its eddies.

'I built my career as an international oceanographer on the *Southern Surveyor*', says Anya proudly. 'The MNF took a risk, not just on a complete unknown, but an unknown young female scientist, and gave me the opportunity to run the ship for three weeks. Suddenly they looked at me and I wasn't just a slightly loud-mouthed young scientist from Western Australia, I was someone who had actually taken ship time from the MNF, transformed it into really really cutting-edge science and published it internationally. It changed my career.'

Martina Doblin

Biological oceanographer

> *'On all sorts of levels we're doing important work for the future of Australia. While that might sound a bit grandiose, it's actually true. And if we don't do it, no one else will.'*

It's just as well Martina Doblin eventually went to sea on *Southern Surveyor*, otherwise she may have packed in her ocean-going research career altogether and her work in discovering a great deal about the nature of microbes in the waters around Australia, not to mention the wider ramifications of those discoveries in terms of our understanding of climate change, may never have taken place.

Her ocean debut, however, was not auspicious. 'As a student I went on a research voyage to Antarctica on the Australian Antarctic Division's icebreaker, *Aurora Australis*', she says. 'I was extremely ill!' It was on this trip that, despite the salving effects of a 'massive injection', Martina pondered that perhaps an ocean-going research scientist wasn't such a wise career choice after all, and that maybe she should stick to the laboratory.

Today, however, after having led three trips on *Southern Surveyor* – the last being 'completely medication-free' – Martina feels she might have shrugged off the not-insignificant motion sickness hoodoo. But it's been a battle. 'If I'm seasick', she tells me, 'it's more than likely others are as well, and I'm well aware of what seasickness does to make decision-making and working in the lab much more difficult.' At my amazement that anyone can even function, let alone carry out meticulous scientific research under such circumstances, Martina says that there are sets of protocols in place for poor suffering scientists and students to fall back on when sick, confused and unable to think: 'Step one – add this to this. Step two – put this here ... etc. Just bring it all back down to the basics.'

Of the many qualities required in scientists to lead a scientific research expedition at sea, Martina suggests two more, unexpected but making perfect sense. 'Empathy and consideration', she says. 'And I have those because I can often feel the worst of it on the boat myself. But in comparison to other vessels, *Southern Surveyor* was the one

I've felt most comfortable on, and which actually changed my view on how well cut out I am for working at sea. Maybe I've relaxed a little bit too', she adds, but she still takes precautions. 'It's embarrassing but I always carry a plastic bag with me, and I'll always suggest we have our meetings nice and low down in the ship, maybe next to the air conditioner!'

Originally from Melbourne, Martina studied in America at the terribly grand-sounding Old Dominion University of Virginia, specialising in toxic algae, estuarine biogeochemistry and many other pursuits which have delivered a PhD and a stunning academic career but which she summarises as 'an interest in examining the lifestyles and nutritional strategies of micro-algae'.

'My last few trips on *Southern Surveyor*', she says, 'involved the incubation of living communities of microbes, examining how much carbon dioxide they utilise and the rate at which they incorporate it into their cell material.'

'And?' I ask, trying to lead with an open-ended question. 'How did they go?' For some reason this elicits a hearty laugh, from which I can only assume they were not unsuccessful.

'When you come out of Sydney and steam eastward', she tells me, 'you see a sudden shift in the water temperature.' This warming is (I feel confident enough by now to guess – correctly, as it turns out) by virtue of the East Australian Current, which flows relatively swiftly down the east coast, from the tropical north to the south. But along with the diversification in temperature there is an equally sharp shift in both the diversity and the types of microbes to be found there, as well as the amount of carbon they are able to take up. This is where it becomes – in the light of global warming – most interesting.

'We know the oceans around Australia are changing', she says. 'They are becoming warmer, and the East Australian Current is moving further southwards. If we can quantify the types of microbes and what they're doing, we can build a better picture of how the oceans' ability to take up CO_2 is also changing. That's become a fundamental question.'

Martina's microbe-hunting voyages on *Southern Surveyor* often involved journeys not of many days into the deep ocean but close to shore, working off river estuaries and river plumes. Although the southern part of Australia lacks the big river systems found in the north of the continent, they are nevertheless significant in contributing to the amount of nutrients available for micro-algal growth, which is an indicator of the general productivity of the ocean. 'Mapping those nutrients and understanding how they're taken up and turned into food gives us a good picture of how land processes affect ocean processes', she says.

The work was carried out by what Martina dubs an 'experimental plan' involving lines of separate observations at increasing distances from the shoreline. 'We called them "cross-shelf observations", taking a slice of the ocean and looking at it to see how the patterns in the microbes reflect the patterns in the changes of the temperature.'

WHAT IS THE EAST AUSTRALIAN CURRENT?

The East Australian Current flows south along the east coast of Australia from near Queensland's Fraser Island. Most of the waters in the East Australian Current split from the coast near Port Macquarie and head for New Zealand. A small part of the current, known as the East Australian Current Extension, continues past Victoria and Tasmania as part of a large 'eddy avenue'.

It is an important feature of the Tasman Sea between Australia and New Zealand. The region features large clockwise and anti-clockwise circulating eddies, up to 200 km across and mixing down hundreds of metres. These eddies slowly move southward to Tasmania over a period of one to two years.

Due to a strengthening East Australian Current, there has been a shift in the temperature and salinity regime ~350 km southward, and a 20% increase in the flow through the Tasman Sea into the Southern Ocean. The waters in this region have warmed by more than 2°C, faster than other parts of the world's oceans.

The changes are thought to be due to a combination of ozone depletion and increasing atmospheric greenhouse gases. Over recent decades the atmospheric circulation has been altering along with changes to a climate system known as the Southern Annual Mode. The changes to the wind patterns that drive the ocean currents in the Pacific Ocean, known as the South Pacific Gyre, cause a strengthening of the East Australian Current, so that the warm tropical waters from the Coral Sea region are forced further south, warming the Tasman Sea.

Regional and global ocean currents in the Australasian region. Source: CSIRO, Wealth from Oceans Flagship and Australian Climate Change Science Program.

The changes are due to a cascade of meteorological processes associated with climate change, in particular the strengthening of westerly winds in the Southern Ocean. Increasing greenhouse gases have strengthened a climate system known as the Southern Annular Mode over recent decades, which has changed the pattern of winds between Australia and South America that control the East Australian Current's southward extent. The easterly flow of water in the Southern Ocean strengthens the South Pacific Gyre, including its western boundary – the East Australian Current. Therefore the tropical waters from the Coral Sea region are transported further south, and warm the Tasman Sea.

Contrasting samples of water were also collected and, along with their resident microbe communities, incubated in a large tank on the deck of the ship. Some samples were taken from the East Australian Current, others from the associated eddies examined extensively by Martina's colleague Iain Suthers.

'It was actually a very large plastic food container inside which were placed several smaller transparent 4 L chambers, illuminated with sunlight, which we can't easily replicate in the lab', Martina says. 'It was like a garden where you could watch things grow, but on the deck of a ship.' The 'fertiliser' for this 'garden' involved the addition of various nutrients to fuel microbe growth: different organisms, like plants, have different requirements.

The experiments were meticulously conducted, replicated three times in each of the smaller chambers according to standard scientific process and observed carefully, to deliver three identical results. Even the influence of *Southern Surveyor*'s steel hull needed to be taken into account, iron being one of the basic microbe nutrients. 'We had to be very particular in the way we sampled the seawater in order to limit the iron contamination which could bias our results.' Martina also staged treatments that included the addition of certain nutrients but not others. 'Some won't grow unless there's a particular nutrient present', she says.

Aside from her research trips, the transit voyages overseen by Martina, where the brief given to her team of students was to look at marine nitrogen and carbon together, hold perhaps some of her most enthusiastic memories of her time on the ship, for reasons that touch on the future nature of science in Australia. 'I've been able to train several female biological oceanographers which has been really satisfying, partly because it's still a pretty male-oriented profession', she says. 'For young female scientists in particular, it's a very empowering thing to be able to do experiments on a big ship, to work at sea and use the equipment. It can be very life-changing.' The voyages were also 'definitely fun'.

Overall, the work carried out on Martina's voyages on *Southern Surveyor* led to a better understanding of the ways different microbes respond to different nutrients. It was, she says, 'quite a big thing that we pulled off'. Nitrogen, the essential nutrient which forms the basis of amino acids and proteins in species from microbes to humans, was found to be in limited quantities in the waters of the East Australian Current and, in a unique observation, both nitrogen and iron were limited in the waters of the adjacent eddy. 'This was the very first observation of that in this region of the world', she tells me, a result partly due to the small number of voyages carried out in the area but also due to the cleaner sampling and experimental practices employed by Martina and her team.

'It was a really pleasing outcome', she says, 'the very first data from this part of the ocean which conclusively showed the ways in which particular microbes respond in different ways to different nutrients.' Her conclusions are approaching publication and she kindly offers to rush me an advance copy, hot off the press, 'But', she adds in a tone of genuine caution, 'I don't know whether you'll understand it.' Her concern, I feel, is probably not ill-placed.

Not one of the scientists I spoke to lacked any sense of the responsibility they had been charged with in working onboard the Marine National Facility, but I sensed Martina bore it with a particular gravity, stressing that all her work, her outcomes, all that she and her team of students and scientists had learned and discovered was about far more than themselves, their careers or building scientific reputations. 'This is a ship that's expensive to run', she emphasises. 'We're privileged to use it, but we have a responsibility to use it well. You're using taxpayers' dollars to undertake this research. It needs to yield something to help us understand the processes in the ocean and how it might assist us in making decisions in responding to changes in the climate. I have that going on in my head all the time.'

She paints a scenario of being approached by a stranger who asks, quite rightly, 'You've just spent a million dollars doing this trip. What have you learned and how's it going to help me?' 'I might answer them by saying "The ocean is worth $44 billion to our economy, we need to manage it wisely so it continues to yield those resources and rewards".' 'On all sorts of levels', she says, 'we're doing important work for the future of Australia. While that might sound a bit grandiose, it's actually true. And if we don't do it, no one else will.'

From a young marine scientist concerned that her future at sea might be curtailed due to seasickness, Martina is now looking forward to time on the new MNF research vessel, *Investigator*. 'When you go to sea with a series of questions, there's always more questions that come to mind as you look through the data.'

'In my wildest dreams', she says, 'I never thought I'd ever be a Chief Scientist, not when I had that experience of being so seasick as a student. But you grow into the roles and what you may have thought impossible becomes eminently achievable in later years.' She has made a pact with herself, just in case. 'From now on I'm only going to the calmer waters, northwards from Sydney', she says. I can't quite tell if she's serious.

Iain Suthers

Biological oceanographer

> *'It was one of those rare moments in science when you realise you've made an amazing discovery.'*

As popular as she seems to have been with pretty much all who sailed in her, the reaction from some quarters when it was proposed in 2001 that *Southern Surveyor* – then attached to CSIRO's Division of Fisheries – replace the ORV *Franklin* as the Marine National Facility research vessel was, according to Iain Suthers, lukewarm to say the least. 'Deep suspicion' are the words he uses. 'How', muttered some within the MNF, 'can we go from a modern purpose-built research vessel to a retired North Sea *fishing trawler*? The Minister is going to have to look at this *very* carefully ...'

Iain admits it did seem a little bizarre at the time but, loved as she was, *Franklin*, which was in operation from 1985 to 2002, simply didn't have the grunt for heavy work. 'She was built for biological and chemical oceanography and some light research trawling', he says, 'but it really didn't have the capabilities for taking the really heavy loads of rocks and deep ocean sampling.' And, jury-rigged though she may have been, *Surveyor*'s lifting capacity from her powerful A-frame was excellent. The geologists on the MNF's Science Advisory Committee campaigned hard for her, and carried the day.

About the time *Surveyor* was transferred from CSIRO's Division of Fisheries to the MNF, Iain was recruited to the MNF's Scientific Advisory Committee, the panel of expert scientists charged with the excruciating task of ranking the applications to use *Southern Surveyor*. It was a role he would fill for thirteen years.

'When it came to the rankings, there was considerable angst involved', he says, quite understandably. Up to 600 days of sailing time were requested by various scientific bodies but only 180 days – including transit times – were funded by the Australian Government. The Committee was a broad-based group constituting geologists, physical and biological oceanographers and geochemists, representing research organisations, universities and institutions.

The Committee's Chair needed to ensure balance, says Iain, 'so that a small but smart and active researcher from, say, the University of Sydney could get equal attention as some of the more highly powered applications from bodies such as (the Australian Government's national geoscience agency) Geoscience Australia'. In fact, he says, the smaller university applicants often fared better, being accustomed to fighting for every research dollar and surviving the blowtorch of the Australian Research Council, which funds a great proportion of Australian research.

'We used three sets of criteria when assessing proposals, and we made sure everyone knew what they were: firstly, the track record of the applicant; second, whether it was good, innovative and novel science; and finally, what would be the national benefit?'

The panel was set up to support the MNF Steering Committee, which includes both scientific and industry representatives. Once the rankings of the applications were agreed, after a process that included external assessments and recommendations from independent referees, the results were brought before the MNF Steering Committee. This Committee took into account factors such as scheduling logistics before making the final decisions. The Steering Committee ultimately decided which of the many competent scientists with undoubtedly worthy requests for *Surveyor*'s sea time would nevertheless soon be receiving a politely worded rejection letter.

The Scientific Advisory Committee's battles were at times heated. 'Yes, there could be a little bit of blood on the floor occasionally', Iain admits with a slightly forced chuckle. One memorable stoush took place with the Chair of the MNF Steering Committee, Craig Johnson from the University of Tasmania, over the somewhat esoteric subject of internal tides off the North-West Shelf. 'I'm a biological oceanographer, and he's an ecologist', says Iain, 'so the two of us were having an argument over something we didn't know much about.'

As a teacher, Iain is very keen to express his belief in the educational importance of the MNF's transit voyages, a Steering Committee initiative, which allowed *Surveyor*'s limited berths to be given to the marine scientists of tomorrow. 'A few years ago, Australia had a love affair with sailing and surfing and fishing', he tells me, 'but in terms of research and industry, it stopped at the edge of the beach.' Increasing the opportunities for students to spend time at sea was vital in bringing Australian marine research from the position where it lagged significantly behind that of Europe and the US. 'We needed to step up', he says, 'and the *Surveyor* was instrumental in this. Money had to be found for the students' airfares, but the MNF provided the ship time and even put on an extra cook!' And, he says, it's worked incredibly well. 'The transit voyages had been vital in encouraging the brightest young scientists to believe that there are postdoctoral and work opportunities in this area of marine science.'

Leading one or two of these transits remains among Iain's most satisfying experiences onboard *Surveyor*. 'I led one from Sydney to Wellington in which I tore

A salp sample collected and brought onboard for analysis. Source: MNF.

my precious plankton trawl net because of the volume of salps we were getting!' Usually salps are seen as not much more than an irritant providing little good science, but Iain is fascinated by these gelatinous zooplankton, which range in size from the tip of a little finger to a shoebox. 'CSIRO researched them in the 1950s but we rediscovered a lot about them, particularly their abundance and phenomenal growth rate.' These astonishing creatures have the ability to increase their size by 5% per hour, and survive between two weeks and three months before being eaten by mackerel and tuna or slowly falling to the seafloor where they collect in vast tonnages. 'They also have tremendous potential for carbon sequestration', he says, 'feeding as they do on the phytoplankton which take in carbon dioxide.'

Although tearing an expensive net was 'quite mind-blowing' (he still has it downstairs and offers to show me) in terms of how many salps reside in the Tasman Sea, Iain was able to factor his findings into wider implications. 'There's a concern that because we've over-harvested fish and unbalanced ecosystems, the Earth is reverting to more gelatinous oceans and an increase in these salps and jellyfish', he says. 'One of the outcomes of that trip was that we went back to the CSIRO's records taken back in 1944, and having used exactly the same type of net, we found there to be no statistical difference in abundance between then and now.' A statistic, he warrants, many may find surprising.

On about half a dozen occasions, Iain managed to divest himself of his responsibilities of teaching and working on the Science Advisory Committee and, as Chief Scientist, escape on *Surveyor* to explore the workings of the ocean, in particular the complexities of the East Australian Current.

It is for what he learned about the dynamics of this major current that Iain considers himself most indebted to *Southern Surveyor* and her crew. He is, I'm discovering, a most thorough thinker – a perfect teacher in fact – and is keen to set the background of his work by stressing the symbiotic nature of the relationship between physical and biological oceanographers.

The warm, clear, nutrient-starved waters of the East Australian Current which flow south from the Coral Sea, he explains, are like a desert running along the primordial coastline beside the nutrient-rich rainforest waters of the continental shelf. Knowing, therefore, where exactly the water you're looking at has come from is vital. 'Whether a particular parcel of water has come down from Fraser Island two days ago, or has upwelled to the surface in the past twenty-four hours, is vital to learning its biological processes', he says. 'Physical oceanographers help us to understand, biologically, the water they're sampling physically.' It is in the mixture of these two waters, Iain discovered, that a paradise exists for larval fish and salps.

To explain further, I am shown CSIRO's Dr David Griffin's *Ocean Current* website, which displays real-time ocean current behaviour based on satellite information. It is a swirling mass of colours and lines emanating from all parts of the coast, with arrows denoting direction and colours for temperature. 'This is the bread and butter for me as a biological oceanographer', Iain explains, then he gives me a crash course in Oceanography 101. I do my best to follow his complex but fascinating exposition of the current and his work. 'Although we call it the East Australian Current, it is really just the western edge of the great South Pacific Gyre which reaches across the South Pacific to the coast of Chile.' This water has spent two years crossing the ocean from South America, being stripped of nutrients, and is crystal clear by the time it reaches the south Coral Sea and is passed down the eastern seaboard of Australia.

He draws my attention to a large anti-clockwise warm-core eddy near Sydney, and another off the town of Eden. Sandwiched between them, however, are more arrows pointing the other way. The sea, I observe to Iain, can look terribly untidy. He laughs at this and points to the screen, 'These are the ones I'm fascinated with – cyclonic circulation clockwise eddies'. It is these bodies of water which indicate where, at around 31° South, the current starts to break up and separate from the Australian coast, heading off towards New Zealand, helped along by winds, and creating in its wake a series of eddies which suddenly allow cold, nutrient-rich water to rush in and surge up onto the continental shelf. This deep water upwelling explains why occasionally, even in the midst of a sweltering summer, the surf off Sydney can be oddly, bitterly cold.

'But it also feeds the phytoplankton, which feeds the zooplankton, which feeds my larval fish and the salps, and ultimately feeds you in the form of that tuna fish sandwich you had for lunch', he says, in that wonderful way scientists have of throwing a large idea out and catching it back again (as well as reducing it to something us mortals can comprehend!).

This is where Iain's work becomes all the more interesting. Wind, he points out, is another major driver of the current as it swings east into the Tasman Sea, and with global warming now generating stronger winds the East Australian Current is accelerating, with unknown long-term consequences. 'This is a very *modern* problem', he says. 'The current has now been intensifying for several decades and we're seeing the changes. The waters around Tasmania, for example, have recorded one of the fastest rises in general regional sea temperatures anywhere in the world: more than 2°C in just eighty years.'

The biological implications of this are tremendous: since the 1980s, an increase in the Tasmanian winter minimum sea temperature above 12°C has seen the arrival of dozens of new fish and other species such as sea urchins which once ventured no further south than Eden. Appearing along the Tasmanian east coast, they have devastated the kelp beds and impacted disastrously on the local southern rock lobster and abalone fisheries. 'These are direct economic consequences of a strengthening East Australian Current', he says. 'And the *Southern Surveyor* opened up my eyes to all of this.'

All the data collected by the equipment deployed into the ocean and by onboard cameras is fed into the computers and monitors in the Operations Room. Source: MNF.

Iain tells me that in research science, the classic 'Eureka' moment of discovery is quite rare. Usually there are slow burns of revelation, drawn out over months, even years, of painstaking and often tedious work. 'But', he adds tantalisingly, 'there was one.' In October 2006, Iain was completing a voyage at the boundary of the East Australian Current and the Tasman Sea, observing how it fluctuates back and forth across the volcanic outpost of Lord Howe Island. At the conclusion of this successful trip, *Surveyor* headed for home, sampling along the edge of the current all the way back to Sydney. 'We were stopping every half day or so', remembers Iain, 'and at one point we received a satellite image showing that an amazing eddy was in the process of forming off Port Stevens.'

A frontal, cold-core, clockwise-moving eddy, it seemed to be 'cupping' some of the productive waters off the continental shelf, and Iain became quietly excited. 'I knew no one had ever sampled one of these eddies', he says. 'They're ephemeral – lasting only two or three weeks.' To Iain's observation, this one seemed to be 'sweeping up and scraping the waters off Foster and Newcastle, enclosing them inside a large gyre.'

Although the end of the voyage was approaching and the clock was counting down to when the *Surveyor* was scheduled to meet the Sydney pilot to guide them into port, Iain was granted some last-minute research time 'as long as we didn't miss the pilot!'

Wasting no time, steaming into the eddy and putting down nets, Iain was amazed at what they brought up. 'They came up absolutely full of small larval fish', he says. 'Sardines, anchovies, mackerel, leatherjacket – I'd never seen anything like it. And it wasn't just one cast of the net, it was all of them, every time we put the them in.'

These eddies, Iain was beginning to realise, were acting as vast offshore nursery grounds for a wide variety of fish and sea life, a notion previously unknown. 'They last for only two or three weeks, but that's enough for larval fish to grow and develop and swim', he says.

To confirm that the environment of the eddy was indeed acting as a nursery, Iain needed to return to the continental shelf and take a comparison sample there – he did so at 4am, just scraping it in before the arrival of the pilot. 'The net came up and we knew we'd hit pay dirt', he says. 'There were lots of zooplankton but hardly a single fish. It was one of those rare moments in science when you realise you've made an amazing discovery.'

Media attention ensued, eager to tell the story of this newly discovered abundance, which, says Iain, 'was pretty good for the old ship'. These eddies have become the focus of Iain's work for the last ten years, and he hopes to keep chasing them well into the future on *Investigator*.

I felt I had learned a great deal about the oceans and currents in a very short time, and Iain's enthusiasm was infectious. More than himself and his own achievements, however, Iain, like many of the marine scientists who had worked aboard *Southern*

Surveyor, was adamant I acknowledge the sterling character of her crew. 'They weren't there just to make up their seaman's salary', he tells me. 'They were there because they knew it stood for Australian marine science.'

Every scientist who stepped onboard was completely reliant on the crew's seamanship, and rarely, if ever, were they let down. 'They're having to cope with a bunch of twelve or so landlubbers – good for nothing but staring at a computer screen – suddenly thrust into the middle of the ocean, and they actually become interested in what you're doing', Iain says.

Something he discovered early was to set up a microscope in the fish laboratory, and just wait for the crew to start coming to see what was under it. 'They'd wander in from the bridge or the engine room, so you'd quickly dump some plankton under the microscope. I don't know if you've ever seen living zooplankton, but it's iridescent and perfect. You almost start to get quasi-religious.'

THE CREW

Ian Taylor

Ship's Master

> *'Some ships seem to have a mind of their own, but in the* Surveyor *we could do anything.'*

For nineteen years, Ian Taylor served as one of *Surveyor*'s Masters, steering her across the waters of the Pacific, Indian, Southern and many more oceans for the betterment of science and the greater understanding of our world and its oceans. Ian's accent gives him away as being from the north of England, Newcastle in fact, and his career at sea began as a young man with an English firm in 1967. Then it was four years on American and other international vessels before he found himself in Australia in 1977. As I speak with Ian, I have images of him hailing from a long line of great Geordie seafarers, and ask if the sea runs in his family. 'Not at all', he says, laughing. 'It's a complete mystery how I drifted in that direction. I still don't know.'

In Australia Ian worked on several ships, including one of the oldest still afloat off the Australian coast, a small 1950s-era converted cement carrier called the *Burwah*. 'An ancient old thing', says Ian, which nevertheless, according to the records, managed to transport a million tons of cement. He progressed as First Mate to the much larger tanker *Esso Gippsland*, staying with her for thirteen years, working up and down the Australian east coast and Tasmania. 'Diesel oil, heating oil, you name it, we delivered it', he says. As First Mate, Ian performed his duties in an exemplary manner, possibly a little too well. When someone was promoted over his head to be given a command, he queried the decision and received a letter from his employer, ship management company ASP, stating that his services were considered too valuable for him to be considered for promotion! 'I kept that letter with me for ages', Ian says. 'People couldn't believe it.'

One day, however, the phone rang and out of the blue an offer was made. Would he consider taking command of the marine research vessel *Southern Surveyor*? 'I was very snobbish about it in the beginning', he says. 'My reaction was, "What, you want me go from a 24 000 ton tanker to a little trawler?"' He grumbled that he'd think it over for a couple of days and get back to them. 'As soon as I put the phone down, though, I started to think about it, then five minutes later called them back and said I'd give it a go. I've never looked back.'

In 1990, Ian's first sight of *Southern Surveyor* was in Hobart, soon after its first voyage, which had been marred by technical trouble requiring it to be towed back into port for repairs. Scheduled to take her out on a series of speed trials, Ian's first trip was essentially a driving lesson. 'She was quite different from normal ships', he noticed, walking onto the bridge for the first time. 'It all looked much less formal. Instead of the traditional ship's telegraph where it was graduated as "dead slow, all ahead full, finished with engines" etc., with lots of ringing of bells and things, here there was just a sort of lever like a car's accelerator.'

There was something unusual about her steering as well. 'There was a wheel, but it was never used', says Ian. 'Instead we steered her by a lever. I wasn't used to that.' Nor had he come across the rather peculiar azimuth thruster. 'The Norwegians put it in. It was powered by a thundering great electric motor on top of the crew room, and you lowered it out through a hole in the bottom.' This appendage was designed as an alternative power source, enabling the ship to move at slow speed in any direction but 'it never really worked. We tried using it a couple of times but all it did was make the ship impossible to steer.' It also increased her draught by an awkward 2–3 m. 'You can imagine being stuck with the damn thing out!' he says.

One feature of *Southern Surveyor*, however, enthused Ian from the very beginning. 'It had these two bow thrusters and two stern thrusters', he says. 'They were fantastic, all new to me, but well, it made *Southern Surveyor* the best-handling ship in the nation. You could do anything with her.'

Aside from taking over the *Surveyor*, Ian found that the job came with a completely new set of responsibilities. 'All I'd had to think about in my old job was taking the ship from A to B and getting the cargo on and off, but it was very different on *Southern Surveyor*.' The domestic running of an entire ship fell under his care, both at sea and in port. 'When we were in port, there was just a skeleton crew of a Master, two IRs (Integrated Ratings) and two engineers', he says. 'You couldn't just sit on your bum, it was quite full-on.' From overseeing the delivery of thousands of tons of oil and gas, Ian had to acquaint himself with such specifics as the ordering of stores, keeping the ship presentable, seeing to the laundry, and even the precise cuts of meat to order for a discerning group of hydrochemists or physical oceanographers! 'And you can't forget anything, either', he says. 'If you're going to sea for twenty-odd days and you don't have enough bread and milk, well, you're in trouble! It was a bit like running a small business where you have to order the stock.' Once he made the grievous mistake of ordering generic teabags, because that's what he drank himself. 'We just about had a mutiny!'

Ian's first voyage was particularly memorable, although perhaps in ways he'd rather forget. Being able to utilise the *Surveyor*'s wonderful manoeuvrability required

Stores are loaded onto *Southern Surveyor* at the CSIRO wharf in Hobart, in preparation for another voyage. Source: MNF.

him to undergo a test to exempt him from the need of a pilot when entering and leaving port. 'It involved taking her up and down the Derwent a few times so they could see I could handle her on my own', he says. 'So when I took her out by myself on a voyage for the first time, the scientists hung a massive yellow L-plate over the stern as we sailed down the Derwent!' Taking the ribbing in his stride, Ian rounded the mouth of the river and sailed east. The sea came up gently, but with a movement that Ian simply wasn't used to. To his horror, he, the new captain, began to feel the grip of seasickness. 'It was terrible', he says, laughing about it now. 'I had my head stuck down a toilet for the first three or four hours of my first voyage. I remember thinking, "I've got to get over this!"'

Unfamiliar motions aside, Ian says the *Southern Surveyor* always felt a safe ship, despite the unaccustomed violence of places like the Southern Ocean. An incident earlier in his career when, as a young man after a late night, he'd committed the

WHO'S ONBOARD THE *SOUTHERN SURVEYOR*?

- **Master** (1) or the captain is ultimately in charge of the safety of the ship and everyone onboard. The Master works with the voyage manager and the Chief Scientist to achieve the research objectives of the voyage.
- **Chief Scientist** (1) is the lead scientist on the voyage. The MNF is available to all scientists employed by an Australian research organisation and their international collaborators. Access is granted on the basis of proposals that are internationally peer-reviewed and independently assessed for science quality and contribution to the national interest. Applications for sea time are called two years in advance with applicants generally knowing twelve months in advance if they have been successful.
- **Voyage manager** (1) represents the Marine National Facility onboard and ensures the safe and effective use of the equipment. They also support the Chief Scientist to achieve the voyage objectives.
- **Scientists** (~10): geologist, oceanographer, biologist, meteorologist (depending on the voyage).
- **Science support staff** (~5): electronics technician, IT personnel, hydrographer, hydrochemist.
- **Chief Engineer** (1) is responsible for the maintenance and smooth operation of the ship. They ensure that the engines, refuelling, reverse osmosis water purification system and sewerage system run smoothly.
- **Chief Mate** (1) is the Master's right-hand man. The ship operates on a twenty-four hour roster and the Master needs to sleep too!
- **Bosun** (1) is in charge of the deck crew.
- **Crew** (14–17): on fishing voyages, more crew were required.
- **Cook** (2): food is available twenty-four hours a day onboard.
- **Shore support** (10), though not physically onboard, are a vital part of the team for logistics, resupply and mobilisation.

cardinal sailor's sin of falling asleep on duty – while seated in the wheelhouse on the bridge, no less – made him swear he'd never sit down there again. Hitherto, he'd managed to remain true to the vow. 'But on *Southern Surveyor*', he says, 'if you didn't sit down in the wheelhouse, you'd be leaving the ship on a stretcher! The movement in heavy seas could be so violent you had to sit down just to be safe.' This was no fault of the ship or her handling, says Ian, but 'just a fact of life on a small ship in a big sea'.

From dozens of voyages, Ian cites his most important skill as the ability to get on with different groups of people. 'I've found with my life at sea, it's really important to get along with everybody. If there are little battles going on it just brings down the level of bonhomie.' He can barely recall when this was the case, but does remember a particularly difficult scientist who insisted there was diesel in the drinking water. 'He just wouldn't give it up', Ian recalls, still with a note of irritation in his voice. 'And I can tell you if there really was diesel in the drinking water, we would all have been pretty sick!'

Once or twice, however, the fracture lines were internal ones within the various groups onboard, particularly the scientists. 'I remember one time when the science team couldn't get along with their own Chief Scientist', he says. 'I spent quite a bit of time in damage control with him, but I can see why the others didn't get along with him. However, I found him all right.'

One aspect of the onboard dynamic that pleased Ian considerably was the level of interest the *Surveyor*'s crew came to take in the science. 'That impressed me more than anything', he says. This general sense of worth, even pride, in the work being undertaken, felt at all levels of the ship, had the effect of 'raising all levels of conversation onboard to somewhere above the navel!' Ian's own scientific education no doubt improved, although he confesses to some of it being 'mind-numbingly boring. Of course you do learn about some areas, but things like CTDs where you have to test a hundred thousand stubbies of water which all look exactly the same – well, no thank you!'

Rarely did the weather determine that the science onboard had to be stopped. 'That was really up to the guys on the deck', says Ian. 'They were at the coalface and I left it up to them to make the call. I could override them to stop them working, but not to make them work if the weather was bad, and that's how it should be. It was a small crew and you really had to be a team. You couldn't have someone up on the bridge who was aloof and untouchable.'

The small size of *Surveyor*'s crew was occasionally illustrated by the navy. 'Sometimes, we did a bit of work with the scientific branch of the navy with mapping and so forth', says Ian. 'Their ships were always smaller than ours and they'd come up alongside and we'd look across and there'd be twelve people on the bridge. Then they'd look over to us and it there was just one person on our bridge – usually me. The navy used to think that was unbelievable.'

As Master, Ian's relationship with the Chief Scientist of the voyage needed to be a close one, from the discussion of the voyage plan before sailing, right through to its conclusion. Were there times when requests were made by those perhaps not familiar with the limitations of ships and he had to just say no? 'A couple of times they'd tell me they needed to have something dropped onto a spot on the ocean floor within 1 m, at which point I'd have to explain to them about currents and things', he says. 'Once or twice they'd give me a route which would have taken us across *land*, which I had to politely decline, but it was rare.'

Overwhelmingly, though, he talks in glowing terms of working alongside the scientists who used the MNF vessel, particularly people like Richard Arculus, who searched for underwater volcanoes. 'A fantastic person to work with', he says. 'He used to bring a whole host of students with him, and I can tell you, they used to worship him. I always look back on those days with a smile.'

Ian remembers *Surveyor* as a happy ship and apparently a dream to handle, a ship which 'flattered the Master' in that it did what it was told. 'It mightn't have been comfortable all of the time, especially when the wind was strong, and as an old ship I suppose we were lucky that more things didn't go wrong with her, but she was so manoeuvrable. Some ships seem to have a mind of their own, but in the *Surveyor* we could do anything.'

Fred Rostron

Chief Engineer

'She was a bit like grandfather's axe.'

Another seafaring Englishman from the north, engineer Fred Rostron hails not too far from Ian Taylor, and offers a perfect summary of just what it was like keeping *Southern Surveyor*'s ageing bones plying the Southern Ocean. 'She was a bit like grandfather's axe', he says in his gravelly Lancashire accent. 'Lasted fifty years but had three new heads and two new handles.'

Coming from a part of the world where big-noting oneself is simply not done, Fred is appropriately modest about his seafaring heritage. 'I had one uncle who was in the Royal Navy', he offers, 'another who was in the Royal Marines. Oh, and there's also the captain of the *Carpathia*', he adds, making me nearly spill my tea. 'Sorry, did you just say the captain of the *Carpathia*?' If so, Fred's maritime pedigree can hardly be more blue-blooded. In 1912, the Cunard liner RMS *Carpathia* attained immortal fame as the only vessel to respond to the distress signals of the *Titanic*, rescuing nearly 700 of her survivors. I suddenly remember from my former obsession with the story of the great ill-fated liner that the *Carpathia's* renowned captain was indeed one Arthur Henry Rostron.

'We *think* he's a relation', Fred cautions. 'All us Rostrons are from the same part of Lancashire and we're all seafarers.' It's certainly good enough for me.

Rain, hail or tumultuous seas, *Southern Surveyor* worked tirelessly in all weathers and Fred hardly got to see any of it. Along with his staff of two fellow engineers and a greaser, Fred was usually confined to his impossibly noisy office deep within the ship's bowels in the labyrinthine engine room. If things were running smoothly, much of his time was spent simply keeping an eye on her throbbing Finnish-manufactured Wartsila VASA diesel engine.

A generally quiet man, Fred, once he gets going, revels in talking about the ship. I struggle at times to keep pace, such as when I ask him to explain the bow and stern thrusters, responsible for giving *Surveyor* much of her renowned manoeuvrability. 'Well, it's quite simple really', he begins. 'You had to put the shaft generator onto the switchboard which was paralleled with one diesel generator, then once you'd split the board you'd put a second diesel generator on to parallel with the first generator, so you had one bow thruster and one stern thruster on the main engine-driven generator, and you had the other bow thruster and the other stern thruster on the Deutz diesels ...' etc. I nodded carefully at everything he said, but I don't think I fooled him for a second. 'Even when things were quiet, there was always something happening to fill the day', he says.

Fred's first encounter with *Surveyor* hurled him straight in at the deep end, an emergency fill-in call he received in 2007. 'She was in Broome and the Chief Engineer's pacemaker started playing up, so they flew him out, and flew me in to fill in.' Upon arrival, Fred found a major problem with a hydraulic pump which steered the Kort nozzle, the cylinder that surrounded the ship's propeller and assisted with steering, particularly at low speeds. It had started to make alarming noises the night before its arrival into Broome, and upon opening it up Fred found that it was 'disastrous ... done for. The ring and the piston had broken and we just couldn't repair it.'

But onboard were a group of scientists to whom research was second nature. Instantly, they plunged into the internet to find ways of either repairing the gear – manufactured by the famous Donkin company in Newcastle-upon-Tyne – or tracking down someone who could tell them about it. That proved not so easy. 'We found the last recorded sale of one of these pumps was to an industrial museum in the UK!' says Fred.

The Kort nozzle or the propeller and rudder of *Southern Surveyor*. Source: MNF.

During the 1980s, working as a registered surveyor with Lloyds of London, Fred had spent time in the Donkins factory, famous since the 19th century for the manufacture of ship steering mechanisms for generations of British ships. 'I recall the steering gears they were making at that time looked completely different to the ones fitted to the *Surveyor*', he says. 'The ones we had were very heavy with big castings and were probably designed in the 1930s.'

A rather jury-built temporary steering system was rigged from spare pumps and bits and pieces, and later in dry dock updated hydraulics were fitted to give *Surveyor* another lease of life.

Later, when Fred joined *Surveyor* permanently, that same Kort nozzle provided an on-going headache, again primarily due to age. 'The welds used to fail so they repaired it and rebuilt it, repaired it and rebuilt it. Some of the welds and parts failed internally so you couldn't even see them. And the stiffener wasn't doing what the stiffener should do, either', he adds as an afterthought (I decided to forgo my burning desire to be fully acquainted with the function of the 'stiffener').

'One time', says Fred, 'a piece of debris had blocked the oil cooler and the rocker gear on number one unit and it overheated. She limped back into Fremantle, and the whole engine had to be pretty much dismantled. It was a big job.'

Problems with *Surveyor*'s variable pitch propeller saw Fred poring over the propeller's drawings, trying to find the cause of the trouble. 'It was two pieces inside the prop connected by a screwed joint, and the nuts had become a bit loose', he says. 'One was moving relative to the other and this was altering the settings on the actual pitch lever. You'd put her in the stop position and she'd still be going astern.'

More straightforward, though no less dramatic, was the problem with the bridge wheelhouse, which at one stage 'damn near fell off'. Erosion had, over time, set in between the bulkhead and the deck, and complaints started to emerge about water leakage into the ship. 'It was found that it was all coming adrift and had to be reattached to the deck', he says.

Every ship afloat needs to be well-balanced in terms of fuel and cargo between Lightweight (empty of stores, equipment and people) and Deadweight (the most it can carry legally) tonnage, or DWT, but the legacy of *Southern Surveyor*'s several incarnations and conversions made managing her a very particular juggling act. As Fred pointed out, she was originally designed as a North Sea trawler, so that the weight of her expended fuel would be replaced by incoming loads of fish, a scenario no longer applicable once she was redesigned as a dive support vessel. 'Then', he says, 'there was all this extra weight of added accommodation, raising her centre of gravity.' This was compensated by filling large tanks in her ballast with concrete, which ate into her fuel-carrying capacity, affecting her range. 'And if we were carrying a good weight of science gear, you couldn't fill the fuel tanks up anyway because you'd put the Plimsoll lines under.'

To undertake maintenance work, *Southern Surveyor* was taken into a dry dock. Source: MNF.

Every Master had their own way of handling the ship, and Fred was a quiet witness to their skills. 'Some captains were better at ship handling than others', he says. 'It's actually more of an art than anything else.' Ian Taylor, he says, had a particularly fine touch, as did Mike Watson. 'They could all manoeuvre her, it's just some had a bit of an edge.' I have little doubt that Fred's opinions of their abilities carried not inconsiderable weight.

Fred is anxious not to paint a grim picture of *Surveyor*, reminding me that a good deal was spent on her in his time and that she was in far better shape when he left her than when he began. His affection for her throughout remained undiminished.

Kel Lewis

Integrated Rating (crew member)

> *'Sometimes you'd have three things going all at once, one after another, all night long.'*

Kel paints a vivid picture of the *Southern Surveyor*'s rear deck where, as one of her crew, he would usually be working for up to twelve hours at a time. 'We did all the heavy stuff', he says, summing it up succinctly. Driving winches and other machinery, deploying and retrieving heavy equipment such as dredges and nets, then 'putting it back in the water and doing it all over again', is how he sees it. Sometimes the scientists would decide on a different procedure from what had been planned, and three separate operations would be conducted in sequence, 'one after another, all night long, twenty-four hours a day. Once it was out there, they hammered it.' The *Surveyor* was a busy ship.

At other times, he'd be up on the bridge doing watches or 'just annoying the captain', as he puts it. I ask how many voyages he completed. 'Gee', he says, momentarily stumped. 'Quite … a lot.' And I believe him.

Like Fred Rostron, Kel takes a while to get going, but once he does, I can't imagine there's much that could stop him. 'A great sea ship', he says. 'I always felt safe on her. It wouldn't have been the same if I hadn't.'

One would hope he felt safe when operating one of her most vital pieces of equipment, her winches. *Surveyor* had six and 'there was usually something pretty expensive on the end of them'. Lowering a CTD 4 km down from the cathouse just above the launching area at the side of the ship required skill and concentration. 'The cable needed to be kept dead straight on the drum', he says. 'But sometimes when the ship was rolling it would become slack, and flick across a groove on the drum. Then you'd have to run the wire back out as the wire could become crushed.' With an average speed of just 60 m a minute wound out and back, deployments sometimes required hours of concentration, constantly keeping an eye on a monitor as the operation played out.

Fortunately Kel was not operating the winch when anything was lost, but occasionally he would bring something up that was totally unexpected. 'One time, in

The back deck of the *Southern Surveyor*, showing her winches. Source: MNF/Rob Beaman.

a rock dredge, I think it was in about three and a half thousand metres. In amongst these rocks, totally unbroken in the net, was a Japanese whisky bottle.' This quite remarkable find stunned the crew who extracted it from the debris gathered in the dredge. 'It blew everyone away', says Kel. 'There wasn't any whisky in it, but it even had its lid on.' The mystery of the untouched Japanese whisky bottle endures, although the bottle itself was quickly snaffled and today resides in parts unknown. 'Probably on somebody's mantelpiece as a souvenir', says Kel.

As delicate as the Japanese bottle might have been, the Sherman sled, a heavy dredge, used to scoop up the larger biological samples from great depths, was built like the famous tank and was a nightmare to handle. 'It was just so heavy and awkward', Kel remembers. 'And if the ship was rolling it could slip around and smash

Crewman Kel Lewis prepares the CTD for deployment off the starboard side of *Southern Surveyor*. Source: MNF/Rob Beaman.

stuff so you had to get it lashed down pretty quickly.' Up it came, full of weed and other samples, which messed up the deck and would have to be hosed off. The scientists, while the dredge was being secured, watched from around a corner for the signal to descend on it and extract what they were looking for. 'We actually had a little roped-off area where they had to wait', says Kel. 'And when it was safe we'd let them come on and do their thing, like kids rushing for lollies, then wash it all down and go again.'

Kel's father had been a ship's engineer and it was all Kel ever wanted to do himself. 'Must be a touch of salt water in the blood', he says. After several years in the merchant marine working on several ships, including the 10 000 ton BHP workhorse *Iron Whyalla*, Kel joined P&O in 2007, sailing with the Australian Antarctic Division's *Aurora Australis*, as well as the Australian Customs vessel *Oceanic Viking*. In 2009, he began work on *Southern Surveyor* for her last four or five years with MNF.

'I used to love going down to the Southern Ocean', he says. 'It was pretty wild but I used to enjoy that. On one voyage we left Sydney, went over to Wellington, then joined the 175° West line at 50° South, and headed north, doing a CTD every thirty miles up to the equator. That was quite a trip.' On a heading that took them across some 'truly terrible' weather, Kel remembers swells of 6–7 m almost until they reached Fiji. 'She was just rolling the whole time', he says. 'We were copping it beam on.'

One of the scientists on this voyage, unused to the adage of always keeping one hand in contact with the ship, was thrown off his feet in the galley by a particularly nasty wave and hurled into a chair, injuring his arm, shoulder and shoulder blade. 'One of the cooks had a bit of experience in sports injuries and was able to strap him up until we got to Tonga', Kel says.

Injuries aside, Kel was usually impressed by the scientists' level of concentration and dedication. 'Once the whistle went and the work started, they just became totally blinkered and focused on what they were doing.' Occasionally, though, you had to let them learn the hard way. 'I'll never forget one kid', he tells me. It was a young student working with Tom Trull on his first voyage on a mooring deployment south of Tasmania. While having a smoko at the 'bus stop' – a small spot adjacent to the CTD launching area – Kel was approached by a confident young man who proceeded to outline all that he intended to achieve over the coming week or so. 'First voyage, is it?', Kel asked after listening quietly for some time to the young man's projected glories. 'Yes, it is', he replied. 'You'll love it', says Kel reassuringly. 'You'll have a great time.'

Later that day, just out of Hobart while the *Surveyor* was still calibrating its instruments over a particularly deep spot in the River Derwent off Opossum Bay, the young man, by then slightly less confident, approached Kel once again. 'Er, Kel, can you tell me something?' Kel looked up. 'What is it, mate?' The young man looked furtively out at the placid waters surrounding the ship. 'Er, does it get any worse than this?' They were still virtually within the suburbs of Hobart and Kel tells me it was so calm, the houses were reflecting, mirror-like, on the surface of the water. Nevertheless, he looked the young man square in the eye and said, 'Mate, this is as bad as it gets. You don't have to worry about a thing.' The lad was reassured.

'The next day', says Kel, 'we had to do a muster. And there he was, this poor kid, holding a bucket under him, spew all through his beard, couldn't move, and that's what he was like for the next five days. Poor bugger. I think he was told to get another career when he got back.'

Another, though more benign, victim of Kel's penchant for ribbing the odd marine scientist or two happened to be the voyage leader, 'Susan', whose particular interest was seabirds. Every morning Susan could be found on deck, binoculars and bird guide in hand, counting birds and writing notes.

The next morning, Kel made a point of approaching Susan with some interesting news. 'Oh, you should've been up this morning, the birds! We had frigates and albatross and … unleaded petrels.' 'Unleaded petrels?' exclaimed the excited woman, with barely a pause. 'I haven't heard of them!' 'Oh yes, they were there all right' said Kel, as nonchalantly as he could.

Later that morning they met again on the bridge, where Susan was furiously thumbing one of the many ornithological guidebooks there for general use. 'Ah, Kel, what was the name of that bird again?' she asked. 'Unleaded petrel', he replied with the straightest of faces, although out of the corner of his eye he noticed the Master, John Barr, scuttling off to suppress a fit of hilarity. Susan remained none the wiser. 'Well, I've been through this book and it's not in here!' she insisted. 'When was it written?' asked Kel. '1975', she answered. 'Well, that explains it', he retorted in a master-stroke of quick thinking. 'It's a new species, you see. Go and ask your colleagues.' The ruse was kept alive far longer than Kel eventually felt comfortable with. Other crew members and colleagues joined the conspiracy, prolonging the poor woman's inevitable embarrassment. Kel eventually called a stop to it, and tracked her down in the control room where she was still haplessly attempting to research the mythical bird. 'Susan, listen to me slowly', he said at last. 'Un-leaded petrel ... ' There was a pause, then a scream of realisation that could be heard in Fiji. In the end, however, there was no harm done, and in gracious surrender she confessed, 'You got me fair and square'.

The weather in the Southern Ocean, however, was rarely fair. One particularly stormy night, Kel and First Mate John Boyes ('Boysey') were together on watch on the bridge as all work on deck had been stopped. Looking out, Kel saw the dark night suddenly seem to darken further. 'Gee, it's all gone black all of a sudden', was all he had time to say before he watched a gigantic wave, taller than the *Surveyor*'s mast, sweep completely over them then simply keep going. 'I just thought, "Unbelievable". She was a strong ship', he says.

A voyage that stays with Kel because of its unusual subject matter was undertaken off Western Australia to learn about crayfish larvae. Hydrochemist Alicia Navidad also worked on this voyage and both she and Kel use the same description for the tiny western rock lobsters which were collected in a fine mesh net trawled at a depth of 20 m or so off the stern. 'Like tiny fine spiders', they concurred. Placed in onboard tanks for examination, Kel says the little creatures became almost invisible. 'You only knew they were there by looking for the shadow they cast on the bottom of the tank. Otherwise you couldn't see them.'

Another tale he tells involves a rather dramatic 'master and commander' type chase, in the wild waters nearly 300 nautical miles south of Hobart, one of the loneliest and most forbidding locations imaginable. A large, recently deployed weather buoy broke free of its 4 km-long mooring cable and began floating into the depths of the Southern Ocean. 'It was massive', says Kel. 'It weighed about 2 tons, and was full of instruments for measuring rain, wind, wave strength, etcetera. We had to deploy another one in its place, then go off and find it.' It's unclear why the mooring failed, but it was decided to track the rogue buoy via its radio beacon, despite the walls of grey-green water that were starting to rise all around the *Surveyor* in those far-off latitudes.

The buoy was tracked to its signal, after which it was up to pairs of eyes with binoculars peering through the gaps in the waves to spot it. 'With those seas, even a big thing like that is hard to find and harder to spot', says Kel. 'You can see it once, and then it's gone again in the waves.' The buoy was eventually retrieved, largely, he says, due to the skills of the Master, Mike Watson, and the *Surveyor*'s renowned

An IMOS Southern Ocean Flux Station (SOFS) mooring, being retrieved from the Southern Ocean. Source: MNF/Max McGuire.

handling. 'Mike got the ship alongside it, and managed to just keep her in the one position going up and down, up and down. How he did it I just do not know. It was the most amazing bit of driving I've ever seen.'

Kel is one of a long list of people who sing the *Surveyor*'s praises, but having also worked in her engine room as a greaser he might be more qualified than most. 'In the time I was on it, there were a couple of little electrical problems and the winches might give trouble, but other than that it was a very reliable old ship. The hull was in as good a condition as they day they built it.' I told him of my tour inside the hull when the *Surveyor* was in dock, and the lasting impression it made. He can only concur. 'Those ribs you saw were only stitch welded', he says. 'Because the steel they used was of such good quality, it never deteriorated.' Once in Fremantle, Kel tells me, a technician was onboard running cables which required holes to be drilled through some of the bulkheads. 'He had a small camera and he could watch it on a little TV, and he flatly refused to believe the ship was forty years old.'

Although it was understood that *Surveyor* was getting past her time, and that her final days were looming, Kel was sad to see her go. 'In October last year I walked down the gangplank for the last time, and it was a genuinely sad moment. There was always something to look forward to when you walked onboard her, you never knew what you were going to be doing but you knew it was always going to be something interesting.'

Kel leaves me with another of those wonderful images of life onboard the *Surveyor* which could fill a book of their own. He was again on bridge watch with Boysey, but this time the Pacific weather was calm. 'It was a beautiful night', he says, 'and I had this little deckchair I used to take out along a walkway out the front of the bridge to sit on. All of a sudden I see in the sky coming across our bow, this green and orange fireball. The first thing I thought was, "Shit, we're under attack!"' Boysey, however, knew exactly what it was – the very rare sight of a satellite returning to Earth and burning up in the atmosphere before disappearing into the ocean. 'He'd seen one before, but it was the most amazing sight. It hit the ocean just a mile or so off our bow. Incredible sight. I'll never forget it.'

John Barr

Ship's Master

> *'Those last six years I spent on* Southern Surveyor *were the best six years of my seagoing career.'*

As the last Master of *Southern Surveyor*, John Barr remembers his final voyage as bittersweet. By the sound of it, one could be forgiven for thinking that, sensing her time with the MNF was coming to an end, the lady decided to make her swansong a memorable one.

'We were picking up some deep sea moorings on one of Tom Trull's infamous rough-weather expeditions, a couple of days' steaming south of Hobart', recalls John. 'Weather-wise, it was horrendous.' Most of the voyage's eleven scheduled days were spent simply waiting for a break in the weather to do the work. 'You've got to get pretty up close and personal to the buoys to hook and lasso them, then bring them onboard', although finding them in the first place sounds like the hard part.

After being tracked to their pinpoint in the ocean, then freed via their acoustic releases from the old train wheels that have been anchoring them to the seafloor for a year or more, the long chain of buoys and floats bob up to reach the surface in the general vicinity of *Southern Surveyor*, and keen sets of eyes scouring the sea are actually required to spot them. 'Tom would always put a finder's fee on the bridge for the person who saw them first', John says. Just as well, as the moorings – particularly after a year or so in the water – weren't easy to see. 'It was hard enough in flat water, but when you're in 6, 7, 8 m swells, it makes it very tricky.'

'They had a strobe light, but after all that time in the water, you couldn't rely on them.' When the shout went up in a 'thar she blows' kind of way, a day-long procedure would begin, the vessel sidling up to the heavy buoy, lashing and securing it then carefully gathering in the 4 km or more of rope, chain and attached instruments. The potential for a seriously tangled mess was great, so John needed to know exactly where along the rope's length he was before starting the winching. 'The sheer length of it could be daunting', he says. 'You don't want to start picking it up in the middle, otherwise you could end up with buoys and rope and all sort of stuff underneath us.

That last trip was a good one from that side of things. We got all of the instruments and floats back, and no one was injured in the process. And just as well, as a lot of people were watching!'

When I spoke with him on a chilly day from his home in Dunedin, New Zealand, John was just shy of celebrating thirty years at sea, having, as a young man, taken a divergent path from his traditional mining family and scored a job at sea onboard tankers and container vessels, working his way up the merchant navy career ladder, plying routes to the UK and beyond. 'It was a great way to see the world as a single man', he says. But after what he diplomatically refers to as a 'dispute over a pay scale', John quit his then employer and 'fluked' it onto *Southern Surveyor* by way of an advertisement in a Wellington newspaper. It was for a position of Mate ('It may have been Second Mate', he says, 'I can't quite remember') for P&O in Australia. Not knowing a thing about the job, or for that matter the ship, John nevertheless applied, was given the position over the phone and in January 2008 headed across 'the ditch' to his new job in a new country. His enthusiasm, however, was short-lived.

'When I arrived in Hobart and pulled up alongside *Southern Surveyor*, I just about went home again', he says. 'It was the smallest, roughest-looking thing I'd ever seen in my life.' Used to far bigger ships, John thought the small blue vessel looked 'particularly sad and unloved' at this stage of her life, and that she looked suspiciously like a fishing vessel, a subject about which he knew nothing.

His curiosity was piqued, however, by the bewildering array of equipment onboard. 'All sorts of gear that was new to me', he says. 'Different kinds of winches and things for deploying moorings and CTDs which I was keen to get stuck into.' As First Mate, he would be under the watchful eye of one of the *Surveyor*'s masters at the time, a man John describes as 'the original British Bulldog', Les Morrow. 'They really don't make them like him any more.'

Soon he would be getting stuck into a very different kind of work from what he'd known. 'The *Surveyor* wasn't like a cargo ship when you were just transporting stuff from place to place', John says. 'You took a lot more ownership of your job, there was much more of a passion about it.' One of the reasons for this was, he says, the presence of the scientists, which struck him as novel. Slowly John began to develop an interest in their work, which by default became his work also. As *Surveyor* put to sea and began heading for the horizons of the Southern and Pacific oceans, John felt his own horizons begin to expand. 'After a while, I'd try and encourage them to come up to the bridge and get involved with what we were doing too', he says. Sometimes the scientists became regular guests, others he met only once. 'But just about all of them were absolutely fabulous people.'

His estimation of the *Southern Surveyor* herself soon also began to rise, particularly on those occasions when the weather became dicey. Once, off Brisbane on a voyage carrying out coring work on subsea landslides at the edge of the continental

The First Mate, John Boyes (left), steers the *Southern Surveyor* as part of the Mawson Centenary Celebrations as Master John Barr (front) and the TasPorts pilot (back) watch. Source: MNF/Sarah Schofield.

shelf, a cyclone came in that *Surveyor* was not able to outrun. 'We couldn't go south and we couldn't go east', he says, 'so our best option was just to nose into it and let it come.'

I ask John to describe what it was like heading into a cyclone in the middle of the ocean. 'Noisy', he says, and laughs. It wasn't so much the seas, he says – *Surveyor* handled the swells well – but the wind, which rose to sixty knots with amazing speed. Then things became 'very dark'. At no time, though, did he feel anything less than confident in *Southern Surveyor*'s ability to handle herself in such conditions. 'She was a beautiful ship as far as weather went', he tells me. 'Just like an old armchair, we used to say. She had beautiful lines and a nice big fat bum and she could ride it all very nicely. I remember it being quite an exciting time.'

Excitement awaited them on the other side of the cyclone, too, as John led the *Surveyor* on one of its most memorable voyages – the discovery of the sunken cargo vessel MV *Limerick*, torpedoed off the coast in 1943, the story of which was vividly recounted to me by geophysicist Tara Martin.

Southern Surveyor's impressive ability to manoeuvre, particularly in open sea, would be tested around Samoa and Fiji with Richard Arculus onboard as Chief Scientist when carrying out rock dredging – as John puts it, 'scraping bits and bobs off the tops of volcanoes'. The description perhaps does not quite do justice to the amount of expertise required. 'At that depth, you have to be quite precise as to where the dredge is going to land, as opposed to where you want it to land', he says. Speed

Since 1988, the *Southern Surveyor* has undertaken 205 voyages, covering around 481 000 nautical miles. Scientists onboard have sampled organisms from 6050 distinct taxa. There have been 4362 CTD casts and the ship has mapped over 315 000 nautical miles with its multi-beam swath mapper. Source: MNF/Richard Arculus.

and currents needed to be taken into account and precisely calculated. 'It was good teamwork, from the bridge to the deck', he adds. 'I stuck to the driving of the boat, but the guys on deck would be driving the winch, having to watch the tensions and getting a feel of working fifteen or twenty-plus tons. They get a sense of what's working and what's not working.'

With a mixture of amusement and admiration, John would watch the eager scientists await the arrival of the next dredge bucket onto the *Surveyor*'s back deck. 'It'd spill out and they'd all charge forward and get stuck into it like kids at a lolly scramble', he says, echoing precisely the observation of Kel Lewis. 'On one trip I remember we did about fifty dredges … of course, sometimes the bloody thing would get stuck.'

In such a case, *Surveyor* would be required to work her utmost to untangle the heavy dredge from the rock prison it had made for itself, often thousands of metres below on the seafloor. 'You'd do a 180, then go back the way you came, trying to find a way, or the right angle, to set it free, always mindful of winching in the cable as you did so, as it could become tangled around the propeller beneath the hull', says John. And if none of that worked? 'You'd have another coffee, another cigarette and try something new.'

Occasionally, when all else failed, it simply became time to 'bring down the hammer and snap the wire', surrendering it to the ocean. John is philosophical. 'If you stick something on the bottom of a place you can't see, sooner or later something's going to get stuck', he says.

After three years as First Mate, John was made Master, his first command. As an old ship, *Surveyor* had issues but 'probably no more than if you had an old car in the garage'. Sometimes 'bits would fall off': one time in dock in Hobart, a diver doing a routine below-the-water-line inspection surfaced with a A4-sized piece of steel which had somehow managed to work its way loose from the Kort nozzle. 'We were planning to go south', says John. 'Instead we turned north towards Sydney to have it fixed.'

At the end of our conversation, John again reflects on those first impressions of *Southern Surveyor* as he pulled up beside her in a taxi to begin his first of many voyages. Dubious then, John tells a very different story today. 'Those last six years I spent on *Southern Surveyor* were the best six years of my seagoing career', he says unsentimentally. 'I had no idea what I was letting myself in for, but I got out of the cab, walked up the gangway and it was probably one of the best things I ever did.' Within her sturdy hull, John learned a good deal about many things, 'about comradeship, about myself, and of course, about how to drive the bloody thing!'

On that final trip, despite the weather and despite being quietly grateful to be back in port, John remembers walking down *Surveyor*'s gangplank, like Kel, for the final time with a sense of melancholy, refusing to quite believe she'd sailed her last trip as the MNF. 'We'd heard that she was going to be retired several times over the years', he says, 'but she never quite had been, so we were still hoping that she'd won another reprieve.'

But it was not to be, and although John is hopeful that one day he will go to sea ('It'd be great to see some of those crazy scientists again') onboard the new *Investigator*, *Surveyor* will be missed. 'She was a great old boat', he says. 'You can be at sea for twenty or thirty years, and jobs like that only come around once in a lifetime.'

Seamus Elder

Engineer

'That ship had character. Real character.'

'If you imagine Ireland being a bit like a little bear sitting up on its backside, I come from about *here*', says Seamus, indicating about half-way up his back. What is it about the Irish? They can turn even a simple explanation of where they come from into a story. Seamus doesn't come from a seafaring family, but perhaps hailing from Dunleary, the harbour town just south of Dublin where the ferry from Britain comes in, inspired him from an early age to go into shipping. 'I didn't actually get away 'til I was 24', he tells me, when he joined Irish Shipping as an electrician, working on tankers and 'cargo liners', sailing to locations like Rotterdam and Montevideo, before finding his way to Australia on 19 February 1968. He remembers the date precisely.

Seamus, I'm learning quickly, has a memory for detail, a fact further brought home when I ask an innocent question about how electricity is 'earthed' on a ship, something which has always mystified me. Half an hour later, after trying to stay on top of an avalanche of information about 'active and neutral isolation' and 'cascading earths', my brain aches and I've learned more about the science of marine electrics than I ever thought possible.

As a fitter and welder with the Marine Board workshops in Hobart, Seamus was in charge of the maintenance of many of its vessels, including no less a craft than the Governor of Tasmania's private launch, on which he was occasionally Engineer. 'I had to wear white overalls which had to be ironed with creases', he says. 'My wife hated it.'

Settling on Hobart's eastern shore with his young family, Seamus' career changed dramatically one January evening in 1975 'when', as he begins with characteristic joviality, 'my wife was in the bath'. Holding a small frog which the cat had caught, and jokingly threatening to place it in the bath to give his wife some company, the shenanigans ceased when 'a tremendous bang was heard and the whole house shuddered'. Racing outside, Seamus looked west towards the city and noticed that some of the lights of the Tasman Bridge were strangely missing and that others seemed to be disappearing beneath the water. He very soon realised that he was witness to perhaps the most infamous night in Tasmania's history, when the 7000 ton bulk carrier *Lake Illawarra* slammed into the Tasman Bridge, knocking a large section of it crashing onto her deck and sending her to the bottom. The large section missing from the bridge effectively cut the city in two. 'I actually watched the *Lake Illawarra* sink', Seamus says.

A few days later, Seamus found himself on the Marine Board's main barge bringing up the cars which had tragically plunged from the bridge into the murky river below. 'The divers they sent in had just come out of training school', he says. 'To be plunged into that thing would have been very hard. We had to lift the cars onto the deck with the people still in them. One couple was still holding hands.' Another car, a little old Morris 1100, he says, had been driven by an elderly woman. 'I can still see her handbag hanging on the choke. It wasn't a pleasant time. I know a couple of the fellas I worked with still suffer the effects of it.'

After several career twists and turns, including time spent as a college technician, a shipping surveyor, operating the Maritime College's training vessels and even a stint in a rescue boat, Seamus was made redundant and contemplated 'a black yawning hole opening up before him'. Putting the upside of unemployment to his wife, Seamus gave her the good news that he could now be around more to help her in the house and garden. 'She just looked at me and said, "Get a job".'

In 2004, Seamus was approached by P&O to train as a casual engineer onboard several of its vessels, including *Southern Surveyor.* Flattered, but feeling he was not strictly qualified, Seamus was reluctant at first, but agreed to a trial period. It was a trial that would last ten years and see him complete around twenty voyages on *Southern Surveyor* alone. 'As a casual engineer', he says, 'I became a sort of mobile unit. If there was a hole to fill on one ship or another, they'd get me to fill it.' In fact, Seamus would regularly clock up more sea time than the permanent engineering staff, with one busy year seeing him complete no fewer than 218 days at sea.

'That ship had character', he says emphatically. 'Real character.' He well remembers the first time he saw her. 'I'd worked with her Chief Engineer. One Sunday, when I was still considering the job, he asked me to come down and look her over, which I did.' A week later, he sailed on *Surveyor* for the first time as Second Engineer for a five-week swing during which he began learning the trade 'rather quickly'.

From the moment Seamus set foot on her deck, *Southern Surveyor* agreed with him. 'It wasn't a big ship, but I remember you could often seem to get to the same place in various ways.' There was also 'a terrific atmosphere on that ship, a certain "aura".' This he puts down, as have many, to the scientists who came and went, sometimes friendly, other times aloof, but always buried deep in their work. 'The places she went and the work that was carried out. That's what made her something special.'

Special or not, *Surveyor* was an old girl by the time Seamus came to work on her. Among his daily list of headaches was her pipework, much of which, though not

The hydraulics room, where all the pumps and pipework to drive the science winches were located. Source: MNF.

exactly original, was nevertheless ageing. Earthing problems with her electrical system occurred regularly and a switchboard, which had been several times converted over the years, presented anyone not familiar with its eccentricities 'a little bit of a problem'.

The Kongsberg dynamic positioning system, in which *Surveyor*'s forward and aft thrusters were locked with the propeller to hold to the same spot in the ocean, 'had been a good one in her day' but was now old enough to have outlived anyone who knew anything about it! Seamus gradually came to grips with its oddities but, he says, 'It'd been rebuilt and repaired that many times, it was a matter of patchwork.'

Nothing gave Seamus trouble, however, like *Surveyor*'s infamous sewerage problems. 'Most ships these days have vacuum sewerage', he informs me. '*Surveyor* just had a flush system which pumped all the sewage into tanks to "de-nastify" it, before sending it out into the sea.' Things went wrong. The three-stage processing tank measured 5 × 4 × 8 feet, and although it was not possible to actually get inside it, Seamus and his fellow engineers had to frequently remove the panel, don protective suits and breathing apparatus and insert their head and shoulders. 'That was enough', he says, and he has the photographs to prove it. 'These were the joys of working in a job like this', he laughs.

One aspect of working onboard *Surveyor* that sounds anything but a joy was the space – or rather lack of space – allotted to the crew members' living quarters. The visiting scientists' cabins were mainly double berths on the lower reaches of the ship

A two-berth science cabin onboard *Southern Surveyor.* Source: MNF/ Rob Beaman.

on the starboard side, Seamus tells me, while the crew and IRs (integrated ratings – crew and deckhands etc.) were single berths on the port side. 'Then you came up a level to all the facilities such as the galley and saloon where people like the bosun (senior crew member) and cook had their cabins', he says. Higher still were the mates and engineers and, above those, the Master and the bridge.

This elevated position, however, afforded little extra in creature comforts, particularly in terms of space. 'Oddly enough, the IRs' cabins were bigger than the engineers' because they'd been installed at a later date. The ones we and the Master had were the original ones. Tiny.' The dimensions of Seamus' living area were 'the length of the bunk, plus a doorway. The width a fraction over 2 m, and the bathroom was exactly 700 mm. I know that because I measured it.'

In the bathroom was a compact shower, a toilet and a 'teeny' washbasin. 'The toilet was angled to fit between it and the shower, so to get to it you had to slide in at an angle between the washbasin and the bulkhead.' These absurdly cramped arrangements might be intolerable for some, but on at least one occasion Seamus saw the advantage. 'I'd become rather ill', he says, 'eaten something that hadn't agreed with me – and found myself sitting on the toilet, while at the same time being violently ill into the washbasin. All very handy!' When we both stop laughing at the image, I suggest it was in fact a very convenient piece of engineering. 'Well, that's what I thought!' he says.

Life at sea is often determined by routine. Seamus' day would begin at 6am with a visit to the engine room, followed by breakfast, then the daily 8am meeting in the control room between all three engineers and usually one engine room IR. Work would be delegated and any outstanding issues addressed. In his career on *Southern Surveyor,* Seamus passed through the three phases of Second, First and finally Chief Engineer, or, as he tells it, 'three phases of responsibility'. These were indeed weighty, with the duty engineer required to be on watch in an 'on call' situation, for twenty-four hours, 8am to 8am. 'His job is to operate, maintain, look after and if need be cosset all the machinery in the engine room, and be in liaison with the Master', he says. 'He also must record all his observations in the log, complete maintenance lists and other reports and, if required, assist the scientists with the ship's facilities in whatever specialised tasks might be required.'

An alarm panel in the control room which monitored systems, temperatures, tank levels and pressures was relayed to the duty engineer's cabin. Again, that infamous sewerage system regularly caused problems. 'Sometimes the automatic pumping system would get a bit sticky', he says. After reaching a certain level, a series of floats would come into play, setting off an alarm and requiring the pump's manual operation to be engaged. To avoid being called out to perform this somewhat onerous task, particularly in the middle of the night, Seamus would give particular attention to the treatment pump and its efficient operation before retiring for the evening. 'The others used to ask me how I managed to be called out so infrequently', he says.

Nor were pumps and blockages the only cause of headaches when it came to the sewerage system. 'The sewage discharge was about twenty feet forward of where the CTD went into the water', he reminds me. This could be disastrous for the samples being taken, and could also contaminate the CTD rosette itself. 'There was a button in the control room', says Seamus, 'which could be used to isolate the sewage pump for a period of twenty minutes.' This was not a problem when the CTD dips were in deep water with durations of several hours, but at shallower depths, with the rosette coming up and back frequently, the operators would keep pressing the button to prevent the sewage from discharging. The holding tank 'only had a certain capacity, and with a full complement of twenty-five or so people onboard', Seamus says, 'it filled up and all had to go somewhere. The next thing you know, you had a spill in the forward hold. Not the most pleasant thing.'

Hopefully it won't be a problem on the *Investigator,* he says. 'The sewage outlet and the CTD station are on opposite sides of the ship. Very sensible!'

Seamus looks back on his years with *Southern Surveyor* with fondness, describing it as a 'high time' in his life. He again refers to the character of the ship, explaining it in terms of the scientists who he met onboard and who broadened his horizons. Once again, Richard Arculus comes in for high praise. 'You could talk volcanoes or rugby with him all day', he says. 'And he was never what you'd call a condescending person. If you expressed an interest in the science, Richard would explain it to you, and make sure you understood, without using scientific jargon.' Most of the visiting scientists, he says, were similarly willing to share their knowledge. 'All you had to do was show a bit of interest and they'd talk willy-nilly with you.'

Seamus remembers in particular the amazing sight from the ship's underwater cameras, and the images of the sea life they revealed. One intriguing picture remains etched on his memory. He can't remember exactly where it was taken but he has a photograph – from one of *Surveyor*'s towed cameras – of a certain object viewed on the seafloor, which for a long time mystified everyone who looked at it. 'It confused everybody as to what exactly it was', he says. 'It was taken in the middle of nowhere, at a spot where the sea bottom is barren. There's nothing there.' After much deliberation, they realised that what they were looking at, kilometres beneath the ocean surface, was an electric beater. 'I reckon it broke in some galley and some ship's cook cursed, saying, "This bloody useless thing!" and tossed it over the side.'

John Boyes ('Boysey')

Chief Mate

'I was the last person to walk down the gangway.'

From those who did one trip to the veterans who sailed with *Surveyor* dozens of times, everyone seemed to know Boysey. Most people probably don't even realise that the name of *Surveyor*'s famous Chief Mate, who was a part of the ship from her arrival at CSIRO Fisheries Division to her very last voyage as the MNF, is in fact John Boyes. To everyone, he is simply 'Boysey'.

'Actually, I was the last person to walk down the gangway', he tells me at his home in Brisbane, where a semi-tamed rainbow lorikeet who has become far too used to a free feed screeches loudly at the veranda screen door for attention.

While not exactly on the water these days, Boysey can certainly smell the ocean from where he lives in semi-retirement, and you sense that he'll always be a part of the sea. He still has that vaguely restless manner of a sailor. During our talks, he never wants to sit down, but rather stands leaning against a bench in his kitchen, frequently glancing outside and reminding me of a man scouring the horizon from the bridge of a ship.

Boysey began his maritime career as a fisherman, then spent time in the merchant navy before being snapped up in 1980 by the newly formed Australian Maritime College in Launceston to become a senior lecturer and run its fisheries training vessel, the *Bluefin*. 'At the time, it was the only independent college for seafarers', he tells me.

The sea, however, beckoned again, and a few years later he assisted with the establishment of the short-lived orange roughy industry off Tasmania, advising on the techniques needed to catch these deep-water fish. A call to come onboard with P&O led him to CSIRO's Fisheries Division research vessel of the time, *Soela*, based in Hobart, with which he stayed just a few months before joining *Southern Surveyor* as Mate. It was to prove a long association.

'She was a hard-working twenty-four hour a day vessel', he says, 'but enjoyable. And a great deal of variety in the work.' As Chief Mate, Boysey describes his job, modestly, as 'running the whole ship'. The Master, he says, made the decisions, but the

First Mate saw that the work was carried out. 'I preferred that operational side to the management side', he says, 'although there seemed to be just as much paperwork!'

Whether on watch on the bridge – Boysey preferred the 4–8 watch, morning and evening – or working on the rear deck, the Chief Mate would make every part of the ship his area, adjusting his hours to the particular tasks required. 'There was always something to do', he says. 'Once you'd finished your watch, you could be straight back down on the deck doing something else.'

Being responsible for the smooth running of the *Surveyor* required Boysey to keep an eye not only on the equipment and its operation but, in the compressed and usually stressed micro-community of the ship, personalities as well. Clashes, though rare, were inevitable, and his skill set required something of the diplomat and occasionally just a touch of the nightclub bouncer. 'It was pretty important that everyone harmonised', he says. 'It wasn't a big ship, so 29 people working together, eating together and living together is pretty tight. Occasionally we might get a bit of a smart aleck crew member who didn't toe the line, but we soon moved them along.'

There was never anything much, maybe an argument that perhaps led to a push and a shove, but Boysey believed it was important to nip such behaviour in the bud. 'I had to move in and break this up, or break that up', he says with a chuckle. 'You try and keep it as discreet as possible, because if it gets out of hand it becomes a really serious matter. You can sense it, when something's wrong or someone's stepping out of line.'

On rare occasions, he says, it was not the crew but the scientists who required intervention. 'Again, you could sense it', he says. 'Tension between say, the two scientists on the same shift. Then there were a couple of times when they were putting too much pressure on the young students and really driving them hard. I had to go and tap them on the shoulder and tell them to slow down.'

The hours, though long, never worried him because 'at sea, you've got nothing else to do anyway. I found it easier just to keep busy rather than come off and sit around and watch a video.' He pauses momentarily before reflecting, 'Actually I'm not really the sort of person that sits down and watches a video anyway!'

The *Surveyor*'s rear deck was where Boysey felt more at home, and his responsibilities there were extensive. Written procedures for every piece of equipment used on the *Surveyor* – how it was maintained, operated, deployed and recovered as well as the vital safety factors – were gone through in 'toolbox' meetings which could be held up to five times a day, 'so even if people weren't experienced, they'd at least be aware of what the procedure was supposed to be', he says.

Boysey would also make sure the correct number of crew were available and that all knew what was needed of them, that the work was carried out on time and that safety procedures were observed. 'Basically, just to make sure everything's up to

Boysey (right) consulting on the deck with Rudy Kloser. Source: MNF.

scratch', he says. 'Because when you put something over the side, all you're interested in doing is getting it back at some stage, so you can't afford to make mistakes.'

With walkie-talkie in hand, communicating constantly with the Master or whoever was in charge on the bridge, he needed to watch everything. 'Even though sometimes it doesn't look like you're doing much', he says, 'your eyes are casting around 360° all the time. If something was going wrong you had to sense it.'

With all the care and procedures in the world in place, however, accidents occasionally happened and, with another of his roles being Chief Medical Officer, Boysey was responsible for the welfare of the injured. Acute seasickness also required action, although just what constituted 'acute' to someone who has never been seasick in his life is hard to imagine. 'You either get it, or you don't', he says. Dehydration, immobilisation, violent coughing, heaving and retching were some of the symptoms displayed by those not so lucky. 'It's a cruel and chronic thing', he continues, 'and there's actually not much we can do when people get that sick. You just make them as comfortable as possible but sometimes, we just had to turn around and put them on shore. If we were too far out though, there was nothing we could do.'

However, a little more than nodding his head in sympathy was required when a somewhat pale electronics engineer, Matt Sherlock, presented himself to the sick bay one afternoon with part of his finger missing! I'd been told the story by Matt himself, but as Boysey was a key player in this bizarre incident, now very much a part of *Southern Surveyor*'s folklore, I was eager to hear it from him.

One of the team watches over the deployment of scientific equipment from the A-frame. Source: MNF.

Boysey had just come off watch and gone to bed when Matt caught his right index finger between a heavy towed instrument which was being recovered, and the cradle it was being lowered into on the back deck. 'The ship was moving around quite a bit and everybody was trying to secure it', recalls Boysey. 'He actually popped the end of his finger off.'

As Matt's work involves small intricate pieces of electronics, Boysey was particularly concerned about his ability to work if missing a digit and, after settling him into the sick bay, issued an urgent order to anyone who could hear him. 'For God's sake, go and find that bloody finger!'

Looking more closely at what was left of Matt's mangled paw, Boysey didn't like what he saw. 'It had sort of peeled off from the knuckle and the end part of the bone was chopped away', he recalls. Shock, he suspects, cloaked some of the pain. 'He was quite conscious so I just spoke silly bloody things to him to take his mind off it.'

The missing piece of Matt's appendage was indeed found, a few metres from where the accident happened on the deck, considerably worse for wear with boot marks, grease and paint chips, and brought post haste to the ship's hospital. 'I remember the cruise leader', says Boysey, 'an American called Tony. He came into the hospital looking for a dinner plate, and there was Matt's finger sitting on it!'

He didn't let Matt see, but Boysey quietly reached for a kidney dish, some distilled water and a new toothbrush, then set about cleaning up the severed digit. A small piece of bone was protruding from the end of the finger, so Boysey decided to try and simply slip the missing bit back on like a glove. 'I was wiggling it around to get it into the right position and thinking, "Gee, I hope I've got his fingernail round the right way!"'

After a few days, with little hope and no expectations, some colour began to find its way into Matt's reattached finger, and by the time they had returned to Hobart it had started to take on some aspect of normality. 'Just as well', says Boysey, 'as I was concerned about gangrene!' The story is, in all seriousness, a tribute to both Matt and Boysey's quick thinking. When Matt showed me the finger in question, it was hard to even see the scar.

On another occasion, this time on a trip to the Gulf of Carpentaria, a party of English science students were onboard for a fisheries expedition. After a haul had been brought onto the deck and taken into the fish laboratory for dissection, one of the English lads decided, for some inexplicable reason, to put his hand down the mouth of a smallish sand shark which, he failed to realise, was not quite dead! The shark, in its death throes, did what all such creatures would be expected to do and bit down hard. The young man's reaction, while understandable, was unfortunate. He wrenched his arm out of the shark's mouth, scraping it against the multiple rows of serrated, razor-sharp teeth in the process. 'It tore his hand to shreds', says Boysey. 'It wasn't a big shark but his arm and every finger were completely shredded.' The student's girlfriend, also a scientist, happened to be standing next to him. 'He was fainting and his girlfriend was screaming. I didn't know which one to treat first!' he says.

While the ship made a beeline for a hospital on a nearby island, Boysey bandaged the arm up as much as he could. A few hours later, a local doctor applied twenty-four stitches into the young man's arm. Boysey doubts whether the student ever repeated the mistake.

It's only after these somewhat gory anecdotes that I notice Boysey is himself missing a digit, something I'm pretty certain he wouldn't have mentioned had I not brought it up. 'Yes, I lost one too', he says, as if noticing it for the first time in a long while. It was indeed on *Southern Surveyor* but not at sea, rather alongside the wharf in Hobart, while he was running new wire on the drum of a trawl winch. 'I was getting a bit impatient, trying to get the job finished, one wire jumped and chopped my finger off', he says. Oddly calm during the incident, he can remember being more concerned for his workmates. 'I was quite aware of what had happened and didn't feel a great deal of pain', he says, 'but I was working with two or three other blokes at the time. I was more worried for them as I thought they were going to faint!'

Medical attention was thankfully close at hand at Hobart's main hospital, but we both reflect on the irony that his finger was unable to be saved while Matt's, in an accident that occurred in the middle of the Southern Ocean, remains a part of him to this day. Not that Boysey's missing finger seems to have slowed him down. 'It took me a while to get used to it', he says, 'but I figured I'd been at sea a long time and this was the only accident I'd ever had. Still, it's funny sometimes when I go to get change out of my pocket.'

Of the enormous range of voyages Boysey helped make happen, it was the ones that harked back to his former fishing days that he enjoyed the most – catching, dissecting and exploring creatures to better understand them, 'biomass, fish stocks, that sort of thing', he says. The more theoretical voyages that gathered endless reams of data for future analysis left him, he admits, a little cold.

Having had such a long association with *Surveyor* Boysey is privy to many of her stories. He offers an intriguing take on one in particular, the the loss of the 'SeaSoar'. Lindsay Pender would tell me the story of how this towed data-gathering instrument had been damaged at sea, but with some *ad hoc* repairs was able to take to the water to resume its job. Eventually, though, said Lindsay (and just about everyone else for that matter), metal fatigue did its job and the SeaSoar snapped at the tow-point then flopped ignominiously to the bottom of the ocean, where it remains to this day. Boysey is not so sure.

Becoming ever so slightly furtive, and with possibly a glance over his shoulder, he proffers an alternative theory.

'I think it hit a bloody French submarine', he says. 'I had a look at that cable. It hadn't just parted, it had broken under stress.' He points out that it would have taken 18 tons to break that cable and as the incident took place near seamounts not too far from the French naval base in New Caledonia, Boysey believes the unsuspecting *Surveyor* was unwittingly incorporated into the exercises of French submariners, who failed to realise she was towing something. It's a wonderful theory, and both of us speculate on a large and mysterious bang that may have startled some French submariners one day beneath the waters of the South Pacific. 'That's the only thing I can put it down to', he muses. Somehow, I hope he's right.

Boysey's given me a wonderful afternoon of remembering his time with *Southern Surveyor*, and as we stand outside his house, with the sea out of sight but close by, I ask him if he still goes to sea. He laughs at that one. 'People ask me why don't you get a boat? Last thing I want to do is get a boat!'

THE GEOSCIENTISTS

Neville Exon

Sedimentary geologist

> *'It was a tsunami. They've been happening along the Australian coast right back through time.'*

'I'm a sedimentary geologist so I'm interested in oil and gas', Neville Exon tells me. And that's just the use he had in mind for *Southern Surveyor* on his first voyage in 2004, under the stewardship of Geoscience Australia. 'We were going to use her for a seismic survey around Lord Howe Rise, off the east coast', says Neville, 'but it turned out we didn't have all the equipment to do it properly.' In a meeting to decide what, therefore, to do with the vessel as they'd already put down the deposit, Neville proposed an idea of his own. 'I think we should go to the Bremer Basin', he said. 'Just use her as a dredging vessel.' Working in this remote area offshore between Albany and Esperance engendered little enthusiasm among Neville's colleagues, especially as dredging was thought to be 'a fairly primitive technique'.

But Neville, who had done a great deal of dredging over the years, had a hunch about the outcrops of the underwater canyons down there. 'No, it's a good idea', he insisted. 'By selective dredging of these outcrops, we can get representative samples from the flat-lying strata going into the continental margin, and build up a stratigraphic column of the layers of strata.' In plain English, very useful information indeed for people interested in searching for petroleum.

Eventually Neville won the day and the voyage took place. 'We did a little bit of seismic work with our gear that didn't work very well', he says, 'but mostly we did dredging.' The result? The discovery of a sequence of Cretaceous rocks with, in his understated words, 'a great deal of interest to industry'.

Nothing was known about this section of ocean floor separating the oceanic and continental crust and no wells had then been drilled, although the presence of petroleum was indicated by tar strandings along the Tasmanian and Victorian coastlines. These occur when petroleum naturally seeps through into the ocean bed and congeals. 'They float on the surface and can be big slabs up to a couple of feet across', Neville says.

The rock dredge is emptied onto the back deck during a voyage to investigate submarine hydrothermal plume activity and petrology near Tonga. Source: MNF.

Although initially greeted with lukewarm interest, this expedition returned an impressive yield. 'We managed to get plenty of nice reservoir rocks', says Neville. 'Also things that could be cap-rocks, and one small but very nice piece of source rock.' These deeply buried indicators of petroleum are 'just what industry is always looking for', says Neville. On the basis of the information gained directly by work carried out on *Southern Surveyor*, the government and Geoscience Australia released parts of the basin, and companies began working exploration leases. 'It was the first expedition I did with *Southern Surveyor*, and it was pretty successful.'

Not that Neville himself was a novice, having already joined many marine expeditions with ships of different nationalities. 'I've been on American vessels, German vessels, Australian, French. The French clearly have the best food!' he adds. One would certainly hope so, although Neville is keen not to denigrate the food onboard the *Surveyor*. 'Once though', he recalls, 'we had a cook who was an ex-submariner. An interesting character but his food tended to fluctuate with his moods.'

Neville began his career not at sea but in the outback, exploring lonely places such as the Bowen and Surat sedimentary basins in Queensland. 'The first time I was in the field was in 1963', he says. 'I spent years out there.' The bulletin he wrote about all aspects of the Surat Basin – subsurface, location of bores etc. – is still regarded as a classic, 'simply because', he says, 'no one's ever gone out there and done it again.'

AUSTRALIAN OFFSHORE OIL AND GAS EXPLORATION

There are few resources with the same impact on the world economy as oil and gas.

Petroleum and its refined products have widespread and important uses – from transport and power generation to manufacturing. The continued availability of petroleum resources has a profound impact on Australia's liquid fuels supply, and the nation's petroleum resources are a major contributor to our economic prosperity.

Australia's offshore oil and gas resources play a critical role in meeting energy needs, and natural gas in particular is a rapidly growing resource export for Australia. The discovery and production of new offshore oil and gas resources requires increased understanding of the geology and environment within our marine territory. It also requires the development of economically viable and environmentally responsible routes to their extraction and transport to market.

Although most of Australia's oil and gas supply currently comes from its marine territory, the deep sea remains relatively unexplored, including several areas that have real oil and gas potential. With only 25% of Australia's marine territories properly mapped, there is still much to learn.

Ongoing research and development is required to provide a better pre-competitive assessment of the potential of little-explored or unexplored areas, and to ensure that the petroleum industry has access to the best technologies and practices to undertake exploration and production activities safely, with minimal impact on the surrounding environment.

Escaping the ennui of the outback, Neville redirected his attention to the sea, completing a PhD in marine geology, not in Australia but Germany. His area of study was the black mud 'dead zones' of the Baltic Sea, with its floor of 'horrible black stinking mudstones full of methane and hydrogen sulfide. Not a good place for general life', he says, or for much else I suggest.

Joining Geoscience Australia's marine group, Neville first went to sea on its vessel the *Rig Seismic*, before leading three voyages on *Surveyor*. One of those was to map the very large but very unknown Kenn Plateau and Mellish Rise, located off Cairns.

Expanding on the geological history of these large underwater plateaus, Neville explains how they once rested against the Australian coast with the continental margin way out to the east into the Pacific, but have been stretched and thinned by the movement of continents to a point where they have sunk into the Earth's crust. 'What was to be found out there?' I ask. 'Well, that was the question!' he tells me, with a burst of enthusiasm, rising from his desk in his office in Canberra's ANU and shuffling through the myriad charts and maps that adorn his walls and hang behind doors. 'Until about sixty million years ago, they were a big part of our continental margin', he says, running a finger along the lines and colours of a large marine chart. 'Something like a hundred thousand square kilometres.' On this expedition they were properly mapped for the first time, revealing their shape and depth, then the seismic and dredging work began. 'Now it's part of the Coral Sea National Park, so it really has no petroleum potential anymore', he adds, sounding almost disappointed. 'But it probably had a little bit.'

This extensive swath mapping was not simply related to natural resources, it also contributed to a biological understanding of the areas. 'Geoscience Australia did the seafloor mapping and CSIRO marine biologists and several universities looked at all the biota', he says. 'What's on the seabed is very important to what lives above it. We were able to map these areas and give the biologists a geological interpretation – telling them what was sand, what was rock and outcrop – to help them put together the bigger picture.' These bio-zonations of the continental margins formed some of the data that contributed to the establishment of the Australian Government's Marine Protected Areas.

I ask Neville to what extent expeditions such as his are journeys of discovery, rather than simply the confirmation of what is already known or at least strongly suspected. He answers carefully. 'What normally happens is that you have some sort of hypothesis, built on your knowledge and understanding of things like the plate tectonics of the area, and often you'd turn out to be completely wrong.' He cites one of his older expeditions to the Exmouth Plateau off northern Western Australia as an example. 'The aim of it was to look at the Jurassic sedimentary rocks that we believed

Igneous petrologist Tracey Crossingham studies a collection of rocks in the laboratory. Source: MNF/Gregg Webb.

were there', he says. 'We started dredging and began to get not Jurassic but Triassic sedimentary rocks which are older, discovering these big reefs which, in Indonesia are oil-bearing.' Suddenly, the Jurassic experts onboard had to become Triassic experts. 'It was a quite a significant discovery and no one knew about it.' There is now known to be a large, but as yet undeveloped, gas deposit in the middle of the Exmouth Plateau. 'You build these hypotheses, then go out and see if you're right or wrong', he says. The surprises are what makes it worthwhile. 'It'd be very boring if you just went out there and found what was expected.'

Neville has given me a list detailing the three voyages he took with *Southern Surveyor,* laid out in excellent detail, each with its own descriptive title. 'Investigation of slumping on the continental slope between Newcastle and Fraser Island', reads his last. Sensing that I may be unsure of what 'slumping' means, he begins to tell me the story of an extraordinary weekend afternoon on Sydney's Manly beach in the 1930s in the middle of a surf carnival. Hundreds were enjoying the waves and the sand on this clear sunny day when those in the water noticed the rip becoming suddenly stronger, dragging them away from the shore. Unlike most rips, however, this one was not broken by the cycle of an incoming wave, and although the water was only waist-high its power began to pull people off their feet. Over two hundred people, at first bewildered but very soon in a state of panic, were sucked out into the ocean and although there were experienced life-savers on duty five swimmers were drowned.

'It was a tsunami', Neville tells me. 'They've been happening along the Australian coast right back through time.' Studies have concluded that possibly hundreds of tsunamis have crashed into the east coast over the past several hundred thousand years, discernible by the telltale deposits of debris and foreign sediment layers in the beds of ancient estuaries. As I am unaware of other causes of tsunamis aside from earthquakes, Neville draws for me an alternative – and fascinating – scenario.

Swath mapping has shown that large sections of the edges of Australia's continental shelf have broken or slid away in vast underwater avalanches into much deeper water, triggering enormous movements of material and water. 'It's like if you had a shovel and pushed a mass of stuff down a slope', says Neville. 'And if it moves very quickly, the water moves down with it, and then it all comes up again, creating a tsunami.' This unconsolidated material can slide down the 5° angle of the continental shelf edge in enormous quantities, up to 200 km^3 at a time. 'Our aim was to go out there, properly map them and take samples of the material immediately above the slump. If you dated that, you'd know when it went.' Some of these spectacular slides occurred up to twenty million years ago, but others were far more recent.

'So the question was', he says, 'were these slides off eastern Australia likely to generate tsunamis?' It's a subject still under investigation, but Neville shows me one part of the seabed where the potential sounds dramatic indeed. Off Queensland's Moreton Bay on the far side of the sand islands of Fraser and North Stradbroke, lies an enormous block of marginal slope – much higher than anything nearby – with, he tells me, a huge crack running right across the top of it. 'You could imagine that if there was a big earthquake, it could go', he says. 'If it went slowly it wouldn't generate a tsunami, but if it moved quickly, it would.' This was not the last I would hear of this alarming structure in my discussions with marine scientists, particularly with geologist Tom Hubble, who referred to it as 'the block'.

This, like all the information gleaned on Neville's expeditions with *Southern Surveyor*, was collected and methodically disseminated. 'I would get people to write up as much information as we could onboard, then after about six months we'd put together a record which might be 150 pages long, which would be available to industry or anyone who was interested', he says. After that, papers and academic journals would follow up the 'more interesting bits', with much permeating out in the general literature and swath mapping information gathered into an accessible national database.

Neville describes his busy ship life onboard *Surveyor* as working with 'a good little team'. With the complexities involved in the sort of work he was overseeing as Chief Scientist, it would need to be. Twelve hours on, twelve hours off, two teams (one beginning at midnight, the other at midday), fitting meals in when you could, hours in laboratories watching over an array of instruments providing a geophysical recording of the seabed below. Usually the sub-bottom profiler was engaged and, if seismic gear was running, a streamer with air guns would be trailing behind the stern. That too would need to be monitored and its data recorded, scientists assessing and modifying their plans as they went along. When dredging, rocks would need to be sorted, examined, described. And then there was the heavy gear, managed by the technical support staff out on the deck. 'You have to have a good relationship with the captain, the mates, the technicians and the seamen', he says. 'And normally, it works pretty well.'

Occasionally, it was *Surveyor*'s worn equipment that disappointed. In 2003, Neville's trip to the Kenn Plateau resulted in everything on the back deck 'suddenly stopping' as a river of ugly brown hydraulic fluid streamed across the deck, the heavy pipes of *Surveyor*'s ageing infrastructure rusting out underneath and failing dramatically. Anxious looks enquired if it was a problem which could be fixed at sea but, with slow shakes of the engineering heads, the boat had to head towards New Caledonia for repairs.

I like to ask all the scientists who worked on *Southern Surveyor* about those wonderful moments of discovery when a piece of evidence fell into the broader puzzle, confirming or perhaps breaking open one of those hypotheses Neville had long held dear. Perhaps there were even moments of sheer surprise? 'Yes, all the time', he assures me. One of his most enduring memories is looking at the reefs that had formed on the Kenn Plateau and Mellish Rise, which began to form twenty million years ago when they were dry land, but are now a kilometre below the surface.

'Some of the reefs which formed around them at the margins had been volcanoes', he tells me. 'We did a lot of dredging to determine when they had formed.' Sometimes what they were bringing up was unexpectedly limestone in character, at other times it was equally unexpectedly volcanic. 'It just depended where you were', he says.

Another aspect which was common was dredging up all sorts of odd organisms from the seabed. 'All these strange little fish with long teeth and lights in front of them', he says. 'If they looked particularly interesting we'd pickle them in whatever we had and pass them on to the biologists.'

One of the biggest surprises, says Neville, occurred when looking for rocks on the South Tasman Rise, a current-swept piece of the continental shelf ~500 km south of Tasmania. 'We started pulling up whale skulls up to 1 m long.' Even more remarkable was how long they had been down there in this whales' graveyard. 'We think they're Pliocene, which means they've been down there about five million years', he says, preserved in part by the coating of manganese which had leached through the seabed over the millennia. 'We would have dredged up about ten of them in this area', he says. 'They were fossilised, but not very much. They actually looked like a fairly modern skull. We had a girl onboard studying nano fossils, looking at tiny microscopic calcite things, when suddenly she has this enormous whale fossil to deal with!'

Despite the meticulous picture Neville enthusiastically paints for me of his work and time onboard *Southern Surveyor*, he describes himself as 'more of an explorer than a details guy. I've been lucky in my career in that there are all these areas we've known "b–" all about and I've been able to go out and find out what's there.'

Robin Beaman

Marine geologist

'It was absolutely amazing. Just one discovery after another.'

You can't quite see the Great Barrier Reef from the window of Robin Beaman's office at James Cook University in Cairns, but it's not far off, and as an ocean geophysicist mapping the seafloor he's seen a great deal of it, particularly on one important *Southern Surveyor* voyage that significantly expanded our knowledge of this iconic natural wonder.

Of the five *Surveyor* trips undertaken by Rob, he counts his first two, in 2006–07, as the most memorable, and has lost none of his sense of excitement when describing what happened on them.

The first, an ocean-floor mapping voyage using the EM300 multi-beam swath mapper (an instrument Rob almost verges on religion when talking about) took him from Darwin to Fiji, across the Coral Sea basin to the edge of the underwater Louisiade Plateau, then further east around the top of New Caledonia to the live volcanic arc of Vanuatu. Here in the subduction zone, volcanoes, says Rob, are 'popping up everywhere. It's very new, very volcanic country where the Australian Plate dives under the Pacific Plate, and you can see it happening in 3D underneath you. It's not just some abstract drawing in a textbook.' He shows me some of the screen captures on his office computer of submarine volcanoes taken by the swath mapper off New Caledonia. As he tracks back and forth with the mouse, the coloured lines and patches denoting depth ('red is shallow, purple is deeper') reveal dramatic images of hitherto unknown structures and seamounts. The violent terrain, with its impossibly sharp slopes and precipices, looks almost alien. 'This is what the raw data looks like when we see it on the ship before it's been cleaned', he tells me with enthusiasm, despite, no doubt, having seen the images many times before.

It is this sense of wonder at the natural world that some years ago led Rob to a dramatic career change. Prior to being a marine scientist, Rob was an officer with the Royal Australian Navy, and working with its early use of multi-beam swath mapping led him to decide that this was where his destiny lay. He resigned his commission,

3D view of a submarine volcanic crater lying on the West Fiji Plate at a depth of 2700 m, seen during a 2006 voyage from Darwin to Fiji. Source: MNF/Robin Beaman.

embarked on a PhD in marine habitats at the University of Tasmania, and has never looked back.

That first voyage also took in the little-known northern part of the Great Barrier Reef, and what was discovered there along its edge took everyone by surprise. 'From a deep water perspective we still don't have a good handle on what lies offshore of the reef in this northern part but for the first time we picked up these massive undersea canyons', he says, pointing out the area on one of the many maps pinned to his wall. 'We had a hint that there were canyons there from some of the old maps but we had no idea how extensive they were. Suddenly what was a flat seafloor now had relief.' Rob would return to these canyons in great detail on subsequent voyages.

I was surprised to learn that the Barrier Reef extends north almost to New Guinea, way beyond the boundary of its World Heritage listing around Thursday Island. Much of it is still a mystery, Rob explains, as he makes sweeps with the back of a whiteboard marker up the Cape York Peninsula on his large wall chart. 'But we do know it's incredibly rich, lots of unmapped coral reefs and really beautiful.'

Rob's second voyage was, he says, the most significant of his career, and brought to life some of the little-known history of the reef as well as offering up a wealth of scientific knowledge that is still being digested today. That three-week voyage was named 'GBR Margins', a title that in no way indicates its importance. 'It was absolutely

3D view of the submarine canyons forming off the northern Great Barrier Reef margin, near Cape York, Queensland, at a depth of 1000 m, seen during a 2006 voyage from Darwin to Suva. Source: MNF/Robin Beaman.

amazing', he says, 'just one discovery after another.' Perhaps the most important of these was the great line of ancient 'drowned' submarine reefs found lying at the edge of the known Great Barrier Reef. Prior to this, a few navy surveys (some of which Rob had a hand in) picked out structures here and there on the reef's edge at depths of about 70 m but, like the canyons, nothing was really known about them.

'There were hints that something was down there', he says, 'but this was when everyone was using the single-beam swath mapper so you never got the full picture.' The promise of what might be awaiting discovery mobilised a formidable array of scientific talent. 'We brought in geology people, robotics people, some oceanographers, biologists and a really good geophysical team.' The voyage's objective was to map a 1500 km stretch between Cooktown and Mackay: no sooner had they turned on the multi-beam swath mapper than it began giving some amazing results.

'As soon as we started mapping in *Southern Surveyor*, we discovered these drowned reefs were everywhere', says Rob. 'One almost continuous chain up to 1000 km long. What we had found was really the *proto* barrier reef, the ancient forefather that was there before the rest of the barrier reef was created.' These were the once-thriving coral reefs which, with the rise of the sea levels, gradually fell beyond the reach of sustainable light. Coring samples have dated them from between 13 000 and 30 000 years ago.

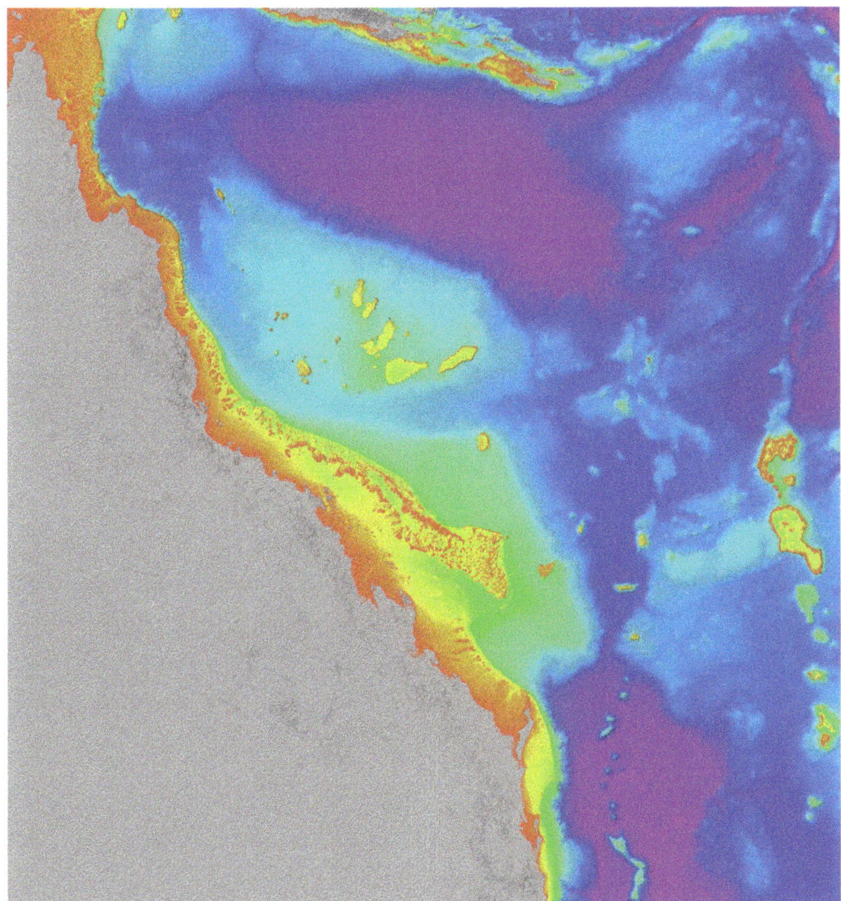

3D view of the north-eastern Australia underwater landscape, encompassing the Coral and Tasman seas, from Papua New Guinea and the Solomon Islands in the north to New Caledonia waters in the east. Source: MNF/Robin Beaman.

He asks me to picture what it would have been like: the sea lapping up against reefs now 50 km offshore and under metres of water, with gum trees, bush and Aboriginal people fishing and hunting at their edges. 'We now have an idea that this was an extensive ancient reef system, possibly the biggest in the world. That was the work done by *Southern Surveyor* in 2007.'

But the reefs were not the only item of interest on this voyage's itinerary. Because of the inherent risks of working in shallow waters, it truly became a round-the-clock operation. 'It was too dangerous to survey these shallow reefs at night', says Rob, 'so at dusk the *Surveyor* would move into the deeper water. Here we were able to get the submarine canyon story, then in the day we'd go back into the shallow parts to the drowned reefs. It was pretty mind-blowing.'

Also used on this memorable voyage was an autonomous underwater vehicle, a 'robot' as Rob calls it, which could make visual follow-ups of the information gathered by the swath mapper. 'As we were mapping, we could say to the robot, "There's the reef, there's a pinnacle, there's a drowned river-bed, go and get pictures of it".' Each robot mission was about six hours long and performed all along the coast. 'We could see the marine life that was growing on the tops of these things, a beautiful veneer of soft coral growing on what were clearly ancient structures.'

The significance of the 2007 voyage, he says, can hardly be overestimated. 'We're still reaping the benefits of it today. We've had millions of dollars of follow-up trips run since, many PhD students, dozens and dozens of papers and we're still drawing on the data.' The information has also been used by the largest marine discovery project in the world, the Integrated Ocean Drilling Program (IODP), which determined that certain sites investigated by *Surveyor* on this voyage were worthy of drilling. At the word 'drilling' I must have looked a little alarmed, as Rob quickly reassures me. 'Not for oil, but to understand what the Great Barrier Reef can tell us about global sea level change.'

In 2010 the IODP drilled along the reef's edge and, as a result, it is now known where the sea level was, what was growing there, and what the temperature and chemistry of the water on the Great Barrier Reef were 20 000 years ago. All this wealth of information is locked in the skeletons of those ancient corals. 'We can look at the coral species', says Rob, 'determine the age of it and peg where the sea level was at the time that coral was created. It's like a library of ocean history, and a very detailed library at that.'

Rob's last cruise in *Surveyor*, in 2013, was an extensive one from Broome to Brisbane, but it wasn't until they reached the Barrier Reef's southern border near Mackay, at a poorly surveyed collection of isolated reefs known as Swain Reefs, that things got really interesting. 'We didn't really know what was down there, but at the shelf break, we found the seafloor was filled with thousands and thousands of these sharp coral pinnacles, about 15–20 m high', he tells me. For at least 100 km, it was discovered, the ancient reefs appear not as one continuous line as happens in the north but in an odd array of what look to me like witches' hats. 'It's strange and we don't know why it's like that, so there's still a lot to discover about the old barrier reef.'

I ask Rob if there's anything he misses about his navy days, or if he ever hankers for being on the bridge and in charge of the ship. No, he says, the work he does today is far more fulfilling. 'You get to ask the deeper questions. In the navy the role was to provide information for safety navigation charts. We weren't then able to go ask, "Hang on, what's that thing we've just mapped and why is it there?" In fact, with the *Southern Surveyor* we did both, as all our information goes to the Australian Hydrographic Office anyway.'

He shows me some more of the wonderful images on his computer, not only from the undersea swath mapper but the working photos taken at sea with an array of

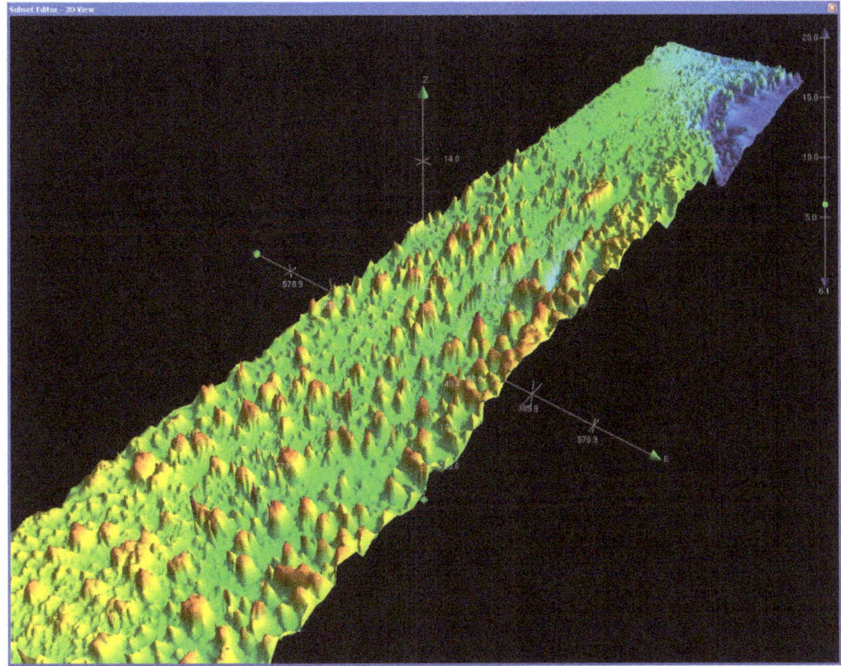

3D view of drowned reef pinnacles lying east of Swain Reefs, southern Great Barrier Reef, at a depth of 100 m, seen during a voyage in 2008 to investigate the vulnerability of ecosystems to acidification. Source: MNF/Robin Beaman.

scientists young and old, watching screens, hauling equipment and otherwise deeply engaged in work. 'I never tire of it', he tells me, as he scrolls through. 'To me it's the closest thing you can get to being an explorer. We're seeing the Earth's surface, the undersea landscape for the first time. It's just so exciting to sit there seeing the ocean floor appearing before your eyes in full 3D colour!'

Rob believes that, particularly with the advent of the multi-beam swath mapper, there's never been a better time to be a marine scientist. 'There's just so much we're discovering about the ocean. Here we have the Great Barrier Reef that everyone knows and loves and millions of people use, and yet not even 50 km away there's this entirely different landscape that we had no idea about until 2007. That was the work of *Southern Surveyor*.'

He tells me he'll miss the *Surveyor*, that is was a sturdy and reliable ship with 'good lines'. 'I loved working on her. She was quite a grand old lady by the end, and you always felt safe on the cruises, no matter what the seas or the conditions were like.' At this he pauses, as we emerge from his office into the corridor where young, earnest-looking science students are making their way past all manner of ocean maps and bewildering tables of marine information displayed on boards. 'You know, I call them cruises but really it's a misnomer', says Rob reflectively. 'People might have an idea that it's all about sitting on the back deck of the thing in banana lounges with cocktails, but really it ain't no cruise. It's full-on, hard work, twenty-four hours a day.' I don't doubt him for a second, and somehow he doesn't strike me as the sitting-around-in-banana-lounges type anyway.

Tom Hubble

Marine geologist

> *'If it were to go and achieve any speed, it could throw a 40–50 m wave back at the coast, which would really be a proper tsunami.'*

'I went to Sydney Grammar where I was an average student, then at university I made it to marginal', says Tom Hubble. So far it's sounding like a career depressingly familiar to my own except that at some stage Tom must have started taking his a little more seriously: today he is Tom Hubble, Associate Professor in Engineering and Environmental Geology at the School of Geosciences, University of Sydney. Over the past decade or so, he's made a name for himself in the rather terrifying realm of landslides along the continental shelf and their potentially cataclysmic consequences, namely tsunamis. Neville Exon's description of coastal tsunamis was still fresh in my mind but Tom, I was to discover, had explored them far more extensively.

Back in those earlier days, Tom suspects his career didn't really start to take shape until working at the Ocean Sciences Institute under the tutelage of Gordon Packham, the 'grand old man' of New South Wales geology. 'If he wanted you to do X and you didn't know how', says Tom, 'he'd just say, "You can read – go and find out!"' He was a teacher who 'appreciated the value of independence'.

After being taken on as a research assistant then occupying various teaching positions, Tom completed a PhD in the intriguing topic of river-bank failure, or collapse, along the Hawkesbury and Nepean rivers, which can happen in times of flood and drought. Using photographic evidence from the 1940s to 1990s, he and his team of students proved that the presence of tree roots along river-banks tended to suppress this sort of failure, then they developed a 'geo-mechanical slope stability model' to provide the reason why. It was 'hideously muddy, and hard, hard work' using a hydraulic ram, picks, shovels, sweat 'and a few guesses. If we'd known how hard it was going to be', he reckons, 'we'd never have done it!'

Not surprisingly, Tom reserves his highest praise for his herculean students who assisted with the months of grunt work of digging up trees and conducting tensile strength tests on root material, while he himself was far away in a classroom. 'For a student, it sounds like a great idea and looks easy', he confesses, 'but they get to the end of it and all swear never to do it again!'

All this toil resulted in a deeper understanding of the mechanics of river-bank failure, but the premise is a simple one. 'A lot of these steep bank slopes have developed with tree roots reinforcing them', says Tom. 'If you remove the trees and saturate them in a flood, you get pretty standard flood failure.' To me at least this would seem obvious, but science is about proof, and Tom's work is about establishing fact from mere conjecture.

Currently, his interest centres on the Murray River, in a project that requires him to spend up to twelve hours a day on, of all things, a houseboat, looking at small landslides along the river-banks, teasing out the causes of the big, almost lethal river-bank failures which occurred during the millennium drought. 'It's some of the most pleasant field work I've ever done', he says, not surprisingly.

With the assistance of a professor of civil engineering from Adelaide, Tom, as both skipper and Chief Scientist, deploys a corer and 'various bits of kit' from the fore and after-decks, probably thinking he's in paradise as he drifts along under the river red gums with neither swell, chop nor seasickness with which to contend. 'The houseboat costs a hundredth of what it does to send the *Surveyor* to sea', he says, an opinion he is indeed qualified to offer – in 2008, Tom's experience with these 'dynamics of river-bank failure' saw him asked to join an expedition to map the 'northern New South Wales and southern Queensland upper continental margin' or, as he puts it a little more colourfully, 'those parts of the continent that descend down to the deep seafloor'.

From river-bank to continental margin failure may seem quite a leap but Tom assures me that it's essentially the same problem, 'only with much more easily identifiable causes. It's the same equations, the same experimental data you need from the soil mechanic's lab, but on a much much grander scale'. From the evidence, these edges of the continental shelf seem to fail in much the same way as do river-banks. Here though, the similarities end, because, really, they shouldn't.

'The continental slope problem is a true paradox', he says. 'It's very hard to understand how those slopes ever fail – really, the materials are too frictional and too strong.' 'Factor of safety' calculations – which compare restoring forces against disturbing forces – rate these submarine structures highly, indicating their integrity to be extremely sound, yet the evidence is that they can and do collapse, and spectacularly so, although thankfully very seldom.

'The only common phenomenon that causes a failure in those circumstances', Tom tells me, 'is severe earthquake shake.' Despite its strength, a powerful earthquake's upward motion causes the frictional interaction of the slope's material to diminish. 'It's like the whole thing is being thrown upwards', he says. The other factor is an earthquake's high-frequency vibrations, which can cause liquefaction of the sediments, when 'it's like the slab above the liquefied layer is on a set of ball-bearings and it just takes off downhill'. The trouble is no one can be absolutely sure because 'no one's been able to replicate this in a large-scale model'.

The edges of the continental shelf, Tom says, begin about 50 km off the coast and drop away down a roughly 3–7° slope, making the seabed plummet from a couple of hundred metres to 4000–5000 m in depth. 'It bounces around a bit with canyons that have been carved out, but we've only mapped it in hi-res above 3000 m and even at that level, not everywhere.' At its best, *Surveyor* could map the seafloor down to 3000 m but Australia's deepest ocean is over 8000 m, so much of our seafloor, he tells me, remains a mystery.

How do we know these failures are actually happening? I'm compelled to ask. Because, he says, of the presence of enormous scars, or scoops, along the steep continental margin of the Tasman Sea, beginning around Bateman's Bay and becoming more frequent as one moves north to around Fraser Island. The biggest of these scoops is the so-called 'Bulli slide' off Ulladulla in New South Wales, which is an amazing 20 km long, 10 km wide and roughly 200 m deep. That represents, I suggest, a great deal of missing material. 'Well, if you relate it to a city', he says, 'it's as if all the northern beaches of Sydney from Manly to Palm Beach and 2–3 km inland all took off downhill one day.' And it's no isolated phenomenon. These scars are 'all over the place', with massive scoops off Yamba, Byron Bay and many other places along the coast.

Dramatic images of destruction notwithstanding, one of the problems, Tom tells me, is that because of the paucity of detailed mapping it isn't known exactly where all this material is. 'We just don't know where it's all gotten to', says Tom, 'except to assume it's gone down onto the lower slope and the abyssal plain below.' That discovery

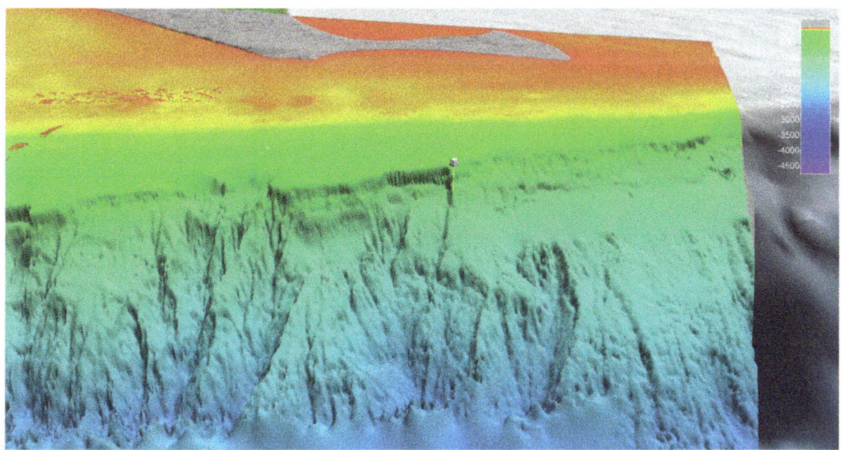

During a 2013 voyage with Tom Hubble, hydrographers onboard mapped the continental shelf off the coast of Yamba in New South Wales, through to Fraser Island in Queensland. At the top centre of this image, Fraser Island is indicated as the grey area. Source: MNF.

must, for the time being, remain the goal of the superior mapping powers of the new vessel *Investigator,* which will be able to map the seafloor to far greater depths. The real – and somewhat sobering – mystery, however, is how often are these failures likely to occur, and what are the consequences if they do?

When I put this question to Tom, he pauses for a moment before saying, hesitantly, 'Well ... I'm not sure I really want to get stuck into this because it's a bit frightening', before immediately proceeding to do exactly that. 'Off Brisbane', he says quietly, drawing me to the edge of my seat, 'there's something we call "the block", a large remnant portion of the margin, about half a kilometre thick and 10 km in length. And there's a tension crack around the back of it. If it were to go and achieve any speed, it could throw a 40–50 m wave back at the coast, which would really be a proper tsunami.' Neville Exon had previously introduced me to this potential scenario, but Tom was somewhat more dramatic! Stradbroke Island, he says, would bear the brunt of this unspeakable catastrophe, but there is a bright side of sorts: 'Looking at that part of the shelf, I think you'd actually run out of water before it could reach too far inland and do too much damage.' Not the greatest of consolations, I suggest, to the nervous residents of Brisbane. 'We're actually going to have a go at modelling that next year', he adds.

Nor is Byron Bay apparently safe, with the 'Byron slide' similarly threatening that seaside village with a 15 m wall of water which would, says Tom, 'overwhelm the dune and take out the town'. However, these and similar events are, he is at considerable pains to point out, very very infrequent. Just how infrequent is still a matter for conjecture. 'We think one of these smaller events happens about once every 10 000 years', he says, and the large tsunamis are estimated to occur but once every 100 000 years.

One of the problems in determining their frequency is the rise in sea levels since the last Ice Age, as the evidence of the havoc they wreak on shore is now drowned under hundreds of metres of water on the continental shelf. The recent tsunamis of our own era have doubtless left indelible images on all who have seen the images of their destruction. As Tom says, 'They pick up everything and dump it on the shore.' But so too does any major cyclonic event, and separating the evidence is problematic. In fact, barely a handful of examples of what 'everyone might agree' to be a tsunami deposit actually exist, says Tom, and of those only one – near Bateman's Bay – has him more or less convinced, and even with this he harbours some doubts. 'It's very difficult to isolate tsunami debris and definitely say it's not a storm', he says. 'It has to be a really thick and extensive deposit that goes 5 or 10 km inland, and there's nothing like that that we know of.'

Tom developed many of his conclusions on two fortnight-long mapping and sampling voyages with *Southern Surveyor,* the first in 2008 led by respected marine

Releasing sections from the corer on the back deck of *Southern Surveyor* on a voyage in 2013 with Tom Hubble. Source: MNF.

geologists Ron Boyd and Jock Keene and another, in 2013, led by himself. The earlier voyage surveyed the shelf between Yamba and Noosa on the New South Wales–Queensland coast, while the second continued the work up to Fraser Island. 'That's when we discovered the frequency of these slides increases as you go north', he says. Despite, on the second voyage, running into the remnants of Tropical Cyclone Oswald, Tom speaks of both as successes. 'We mapped the slope between Noosa and the north of Fraser Island in great detail, then did a program of samples along the slides that had been identified on the 2008 trip.'

The mapping was done at night, while the sampling occupied the day shift. Using Geoscience Australia's 6 m gravity corer, known lovingly as 'Thomas the Tank Engine',

Surveyor's rear A-frame and winches drove the corer's 1000 kg of lead into the seafloor at roughly 12 km an hour – 'the pace of a medium jog' – allowing a vertically hanging pipe to ram into the bottom to capture some of its content.

To make the most of these precious shelf samples (the consistency of which Tom describes as 'more workable than Playdoh but not exactly goo'), the captured muddy sections were brought up to the *Surveyor*'s rear deck, released from the corer and cut off in manageable, pipe-shaped 90 mm lengths. They were then photographed and a preliminary sub-sampling was taken for carbon dating.

A strict protocol guided the journey of these precious clues to the make-up of the continental slopes from the back deck and into storage. 'There's a procedure of labelling', Tom tells me, 'but sometimes the indelible ink washes off, so we'd also have to engrave them.' Information of the cruise and site number, location, core piece, date and sometimes even the Chief Scientist's name were engraved into either side of the cylindrical barrel of the core once it reached the ship's wet laboratory, before being refrigerated. When Tom next saw it, it would be in a different laboratory somewhere onshore, often many months later and under very different circumstances.

Of course, Tom was never alone in performing these tasks and, as he says, being on the *Surveyor* was in many ways easier than working on shore. 'It's a wonderful thing being at sea. As an academic, I usually have only one person helping me deliver my prac classes. On a ship, you've got ten or eleven scientists or students or colleagues all giving you a hand across two shifts.'

Tom is anxious to go back to sea to answer the 'Where does this stuff get to?' half of the slope question paradox. Not content with simply knowing it must be down there somewhere, he says much more information could be gleaned by understanding precisely *where* it has ended up. 'That gives us a sense of how far this material travelled, whether it disaggregates as it moves down the slope or remains in big solid lumps', he says. 'There's evidence to suggest that both those styles of failure occur. I really want to get back out there on the *Investigator* to map the lower slope and to find out where this stuff gets to and in what form.'

Many of Tom's conclusions are still being worked up for future publication. The question of what exactly causes these submarine landslides from the heights of the continental shelf and what the consequences might be is, he says, one of the holy grails of marine engineering science. One suggestion might be the very cold polyna currents generated in Antarctica, which pour into the abyssal plain and force flows up towards the equator. 'One of those flows, I suspect very strongly, might be eroding the toes of the slopes, preconditioning them to failure.' Then again, he adds, it might also be 'the occasional large earthquake'. Does he, I wonder, spend hours contemplating such terrors and catastrophes? 'It doesn't preoccupy me', he says. 'Rather, I think I'm a scientist because I'm curious.'

Nevertheless, with a little prompting, Tom divulges what he believes such a shelf slide would entail, and suspects the terrible images of the Boxing Day and Japanese tsunamis are actually misleading. 'Those were much larger in scale and more widespread', he tells me. 'These events would be far more localised, but bigger.' As an example he refers to the Scotch Cap disaster of 1946 when a 7.4 magnitude earthquake struck south of the Aleutian Islands in Alaska and 'shook something loose', sending a 30 m wave crashing into Unimak Island, destroying a newly completed lighthouse as well as much of the cliff that was supporting it, with severe flooding as far away as Hawaii. 'If you were near the coastline and an earthquake of that magnitude occurred', he ponders, 'well, you might want to get to higher ground in about fifteen minutes'. Sound advice.

Patrick De Deckker

Geologist/zoologist

> *'People think the oceans are getting 1° warmer everywhere. It's not entirely correct.'*

Some scientists speak of their oceanic explorations as voyages, others simply as trips. The French on the other hand, with whom Patrick De Deckker spent some time in the pursuit of his extensive study of ocean sediments, refer to them with wonderful Gallic haughtiness as 'campaigns'. The boldness of this, I confess, appeals to me, and despite not being at all French, but Belgian, I suspect it also does to Patrick.

Beginning his career not at sea but in the desert, Patrick spent a good deal of time traipsing across the barren, billiard-table surfaces of the great salt lakes of the Australian outback, attempting to discover their climate history. Kati Thanda–Lake Eyre, Lake Buchanan, Lake Frome – Patrick got to know them all. The trouble with salt lakes, however, is that they are for the most part particularly lacking in water, which makes trying to look at their sediments deposited over the millennia somewhat difficult. 'Today the lakes are dry, the sediment has been blown away', he tells me. 'The history is incomplete.'

Far better, he thought, to explore sediments where they are preserved – the bottom of the sea. 'That's why I dedicated twenty-five years of investigation to the sediments of the open ocean', he says.

Coming to Australia in 1970, Patrick completed degrees in earth sciences and geology, as well as a PhD in zoology. 'So, I'm actually a geologist–zoologist.' This, as it happens, is an ideal combination of skills when unlocking the secrets of sediments. 'After plankton dies', he tells me, 'their skeletons are preserved on the seafloor. Then there's the wind-blown sediments and muds from rivers discharged at sea, so together I can have a record of what happened at sea as well as on the land.'

Sediment, I'm learning, holds a great deal of history, particularly in a largely flat continent such as Australia where erosion has been minimal. One metre of sediment, I'm told, can represent an astonishing 100 000 years of deposition at sea. 'They're the archives of the seafloor', says Patrick, and indeed it seems as if he can read them like a book. 'In a pile of sediment, I can tell through various plankton species if the sea was warm or cold; I can find wind-blown pollen which can tell me what vegetation was on the land; I can see by the amount of river discharges if it was a wet phase or an arid phase.'

Even the history of the East Australian Current, which brings the warm waters of the Coral Sea down the east coast to Tasmania, can be detected. 'We can tell if it was always functioning, or if it was stronger or weaker. We can also read evidence of La Niña and El Niño periods.' The ideal depth to work at, says Patrick, is at 1000 m, where the ancient Ice Age deposits have been relatively well preserved. Where this depth is found varies depending where you are looking and the variations in the continental shelf. 'Places like south of Adelaide it's about 200 km offshore, but east of Sydney, it can be just a couple of kilometres away.'

Twenty thousand years ago during the last great Ice Age, a large part of the North American continent as well as the British Isles and Scandinavia were covered in places by glaciers up to 4 km thick. In Australia, where sea levels were 120 m below what they are today, it was possible to walk from New Guinea to Tasmania. This is the period Patrick De Deckker set about to unearth on *Southern Surveyor*, concentrating his efforts off South Australia, following the ancient course of Australia's greatest river, the Murray.

On earlier voyages with the French research ship *Marion Dufresne* which, aside from its scientific worth, must surely rank, with its sharp black and white livery, as one of the most elegant marine exploration vessels afloat ('and let's not talk about its *restaurant*!'), Patrick began by mapping some of the great undersea canyons off the Murray's mouth. These 'Murray Canyons', he says, have 'phenomenal features, deeper than the Grand Canyon.'

In 2005 he returned to these same waters, this time with *Southern Surveyor*, to learn more about the ancient history of the Murray River and to map its long-submerged course of bends, tunnels and curves, carved across the face of the 200 km-wide Lacepede Shelf. 'Twenty thousand years ago', Patrick tells me, just to put me in the picture, 'you could walk 200 km from the mouth of the Murray, south to the edge of the shelf, somewhere east of Kangaroo Island.' Patrick undertook 1700 km of seismic surveying along the courses and meanders of the old river, now far below the surface of the sea. 'The course we took made it look as if the captain was drunk', he tells me, but some of the mapping images obtained of the old river-bed are indeed spectacular.

Having gained an understanding of the layout of the river, Patrick was eager to find out what sediments lay in its ancient meanders, and for this purpose a corer was needed. This device, used widely on many of *Surveyor*'s expeditions, weighed roughly a ton and was in essence a 6 m long stainless steel pipe, open at one end and with a piston at the other. When jammed into the bottom, the 10 cm wide pipe filled up and was then closed off with a diaphragm. The sediment sample was held by the suction of the retracting piston before being raised to the surface. On this occasion, however, it proved not to be the right tool for the job. 'There was a lot of sand on top and it was just bouncing off', recalls Patrick. Taking the information already gained about the river-beds, the expedition decided to wait a year until they could tackle the Lacepede Shelf again.

On the next voyage, conducted in 2007, a new piece of equipment was used, courtesy of Geoscience Australia, which even sent along its own technician to oversee its use. Enter the 'vibro-corer', essentially an ordinary corer but 'a little bit like a jackhammer', with a mechanism enabling it to vibrate its pipe down into the ancient river-bed to capture a small parcel of its history.

At this point my imagination wants to propel me along these deep subsea river-beds, but I'm struggling to visualise this surreal terrain. What does the course of a drowned river-bed look like? Would it be obvious to anyone down there what they were looking at?

'Have you ever been to Murray Bridge?' he asks. I have, in the past year or so. I remember the terrain well, as the historic iron bridges mark the river on a sweeping bend, from where it continues its course through dramatic walls of near-vertical cliffs. 'That's what we found on the shelf', he tells me. 'It enabled us to determine its ancient size and dimensions.' Remarkably, Patrick believes that until a century ago, when the Murray's flow was impounded and reduced to the often struggling watercourse it is today, it emptied robustly into the sea, following these ancient bends, a river flowing under the sea. 'It's not too surprising when you consider the Amazon, which continues to travel underwater for another 1000 km out from the coast.'

This ten-day voyage was a remarkable one and achieved at least one record. 'We wanted to understand how the canyons at the edge of the Lacepede Shelf formed', says Patrick, 'so we lowered a corer down to the bottom of one of them.' The gargantuan dimensions of these canyons are a little hard to comprehend, some being 2 km across and more than 5 km deep. 'We lowered a corer down to 5300 m', he says. 'To the absolute limit of the cable. I think it's a record. No one's ever retrieved a core in the ocean that deep offshore Australia.'

What, I wondered, does ancient river sediment look like when it's brought on deck, having just been ripped from the seabed nearly 5.5 km below? A sloppy pile of sludge? I was in for a surprise. The 'yogurty' top 50–70 cm is in fact a soup of calcium carbonate crystals, water and the skeletal remains of algae and plankton, but then it becomes firm and malleable, like a variegated sausage. 'Remember you've had several kilometres of water pushing down on top of it so it's quite compressed and layered', says Patrick. 'Sometimes, if there's a lot of organic matter it really stinks, like rotten eggs.'

There was, however, no time to examine these samples as they arrived onboard the *Surveyor* from the stygian depths. Instead the cores were immediately cut into 1 m lengths, carefully catalogued, cool-stored and, when back on shore, placed in a special cool room at the Australian National University (ANU) at exactly 4° to

COMMON RESEARCH TECHNIQUES USED ONBOARD *SOUTHERN SURVEYOR*	
Gondola 	Mounted 1 m below the hull of the ship is a scientific gondola containing sonar equipment to map the seafloor (swath mapping). This allowed scientists to map the seafloor to depths of around 3000 m. It works by transmitting and receiving a sonar (acoustic) signal that travels from the ship to the seafloor and back. The time it takes to bounce back to the receivers provides a measurement of depth. This system revolutionised the way science was done onboard the MNF vessel, enabling oceanographers to pinpoint locations for deploying moorings, uncovering the mysteries of the seafloor to geoscientists in search of resources and providing accurate images of habitats for biologists.
Rock dredge 	The rock dredge has a 12 mm thick galvanised steel opening that is ~90 × 30 cm; attached to this is a chain bag, made from 6 mm galvanised steel links. At the end of the bag are two pipes 20 × 50 cm that collect smaller particles. While it might be the least technical piece of equipment used onboard *Southern Surveyor*, its role is critical in confirming the location of mineral resources and discovering the movement of tectonic plates over millions of years. Rock samples collected onboard *Southern Surveyor* discovered how Australia, India and Antarctica parted ways 130 million years ago and allowed final confirmation of the location of billions of dollars worth of oil and gas reserves within Australia's Exclusive Economic Zone.
Smith McIntyre Grab 	The Smith McIntyre Grab (or Smith Mac Grab) is a stainless steel sediment collection device that collects mud and sand samples. It has two jaws which are spring-loaded onboard; when it comes in contact with the seafloor, the jaws are automatically triggered, scooping up the top layers of sediment. This allows scientists to work out what animals live in the top layers of the seafloor and to determine the mineral composition of the seabed.
EZ Net 	The EZ Net consists of a large frame, ten fine-mesh nets and trigger system that is connected to the ship's operations room via a fibre optic cable. The ten nets collect samples of phytoplankton, zooplankton and small fish down to 1000 m. Scientists receive data live from the EZ Net into the operations room, which they use to determine when to open and close the nets to collect samples. The EZ Net also can be fitted with cameras, optical plankton counters and CTDs.

CTDs 	The CTD (conductivity, temperature, depth) device collects a wide range of oceanographic data. As it is lowered into the ocean, sensors also monitor ocean properties such as light levels, pressure and turbidity. As it returns to the ship, water sampling bottles are closed remotely at different depths by scientists in the operations room onboard. The water is analysed for nutrients such as nitrate, phosphate and silicate and for trace elements such as iron, which is important for plankton growth. CTD data has allowed scientists to discover the southward movement of the East Australian Current, which is changing ecosystems, bringing warmer water species further south.
Moorings 	A mooring is a long anchored line of scientific equipment and floats, deployed to collect data from the ocean for one to two years. Mooring lines are usually deployed with the anchor last, and a parachute is used to create a gentle descent. The top of a **subsurface mooring** is typically 20 m below the surface, to avoid any impact from ocean waves and ships. Subsurface moorings measure things such as temperature and conductivity (saltiness) at the same depth, across an ocean current. The speed of the current can be measured using sound waves. **Surface moorings** collect ocean, meteorological and atmospheric data such as wind speed, wave height and temperature. Below the surface it also records data such as current, conductivity and temperature. Scientists use mooring data to measure changes over time, and to learn more about important ocean currents such as the Leeuwin Current.
SeaSoar 	Scientific instruments can be towed behind the ship and controlled by scientists onboard via a fibre optic cable. One of the key pieces of equipment used predominantly by oceanographers was the SeaSoar. The SeaSoar is a type of towed ocean-profiling robot that can collect data from the surface down to 600 m. It is fitted with sensors to measure temperature, oxygen levels, salinity and turbidity. Turbidity (the cloudiness of the water) is important as light is needed for photosynthesis to occur in phytoplankton, which are the very beginning of the food chain.
Argo 	An Argo float is a robot shaped like a pencil or a rocket and it contains some of the most advanced scientific equipment on Earth. It is deployed into the ocean where it will dive to 1 km below the surface, drift for a while then dive to 2 km depth and then collect data all the way to the surface. Once it has surfaced, it will send data back to scientists via satellite. It will continue to do this for up to eight years. As part of the International Argo Project there are over 3000 Argo floats currently in the world's oceans, constantly relaying data about ocean movement and health to scientists who share the data freely.

Fishing

In its time as the CSIRO Fisheries Division vessel, every time *Southern Surveyor* went out to sea there were new discoveries and the deeper the water, the more species found that were new to science. Two examples of species discovered by scientists onboard *Southern Surveyor* are:

- the eyebrow wedgefish (*Rhynchobatus palpebratus*), which was officially named in 2008. The species grows to about 1 m and is found at depths of 5–60 m in tropical waters off western and northern Australia and off Thailand
- the cockatoo handfish (*Pezichthys amplispinus*), known from only seven specimens collected from southern New South Wales and eastern Victoria at 74–121 m depth; six were collected by *Southern Surveyor* in 1996. Fourteen species of handfish can be found in southern and eastern Australia; they occur nowhere else in the world.

Both samples are now part of the Australian National Fish Collection which is managed by CSIRO in Hobart.

Fisheries research data was collected and used to provide advice to those who manage Australia's fisheries. Scientists work in collaboration with industry; for example, deeply towed echo sounders were used and egg production research was undertaken to develop population estimates for commercial species such as blue grenadier and orange roughy. Also, in waters near Macquarie Island, scientists studied the Patagonian toothfish fishery and its ecosystem, which led to a greater understanding of the species.

replicate the temperature of the ocean floor. 'We have 900 m of sediment cores stored here', says Patrick. 'The business manager is not happy as they're taking up so much space. He wants me to get rid of them!'

Examining such a large amount of samples takes time, and although many of Patrick's findings about the old Murray have been published there is still a great deal of work to be done. It will, he says, be years before all the sediment samples are examined and their secrets revealed. However, he can say that from what he has learned, there was not one course of the Murray but several, over different phases of sea level change. 'We think this is important because there were most likely people living on the edge of the Murray. We also think there could have been a very large lake, but we haven't found it yet.'

Patrick tells me he also wanted to know what lived in these waters. 'In the same way you have tropical plants and arid plants, so it is the same with plankton: some inhabit tropical oceans, some live in the southern oceans and they're quite different. Currents bring tropical plankton during warm phases and I spent a lot of time collecting plankton from different regions.'

Two shorter voyages undertaken by Patrick on *Southern Surveyor* in 2011 concentrated on quite a different area, and one with particular contemporary resonance, the changing temperature of our oceans. These were transit voyages, where the logistical necessities of *Surveyor*'s schedule were utilised to the full for scientific work. 'They were voyages from Hobart to Brisbane, and Fremantle to Hobart', says Patrick. 'We only had twenty-four hours of research time during the course of the transit, but we achieved a lot.'

The Bureau of Meteorology produces maps of Australia showing temperature changes over decades. Patrick believes he has a way of going back way, way further. 'People are talking a great deal about global warming but we only really know about the last fifty years', he tells me. 'We have a technique where we can look at the chemical composition of some plankton, which relates to sea surface temperature going back 500 years.'

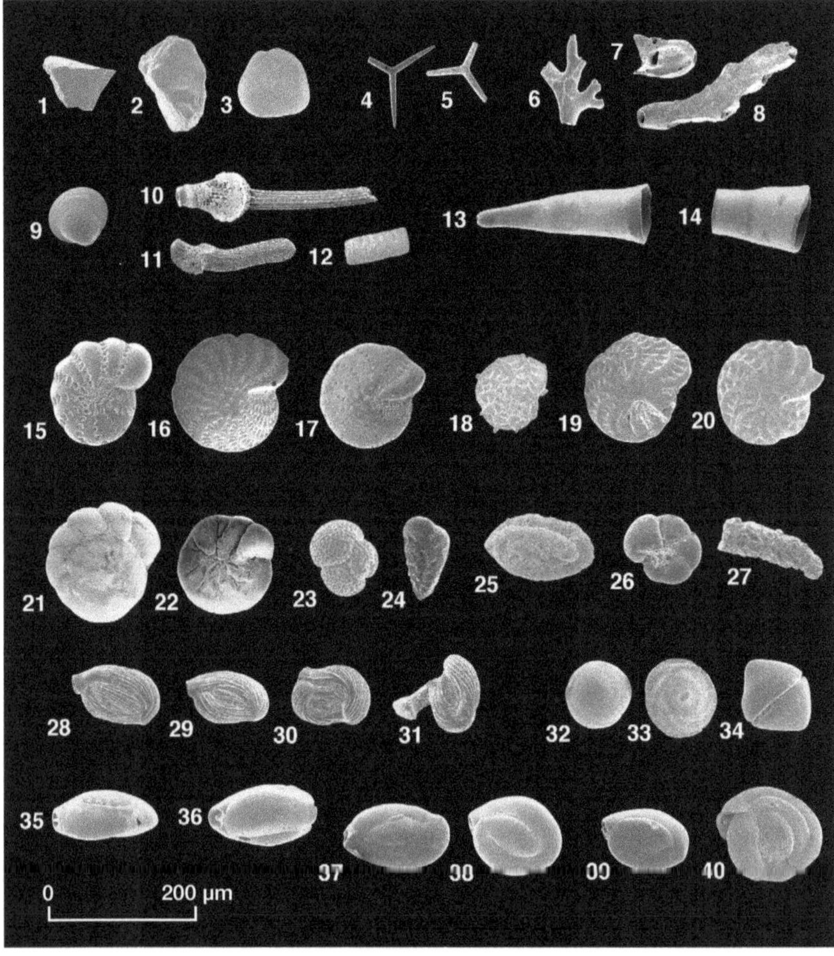

Scanning electron microscope image showing microfossils and other materials from sediment cores. These findings are used to analyse the history of a region. Source: Graham Nash.

The procedure is highly complicated. It involves a coring device called a multi-corer, a heavy 2 m tall tripod surrounding not one but eight small clear plastic tubes, around 60 cm long. On the seafloor, these are pushed into the soft yogurty substance at the top of the seabed Patrick described earlier. 'We collected 30 cm of this yogurty substance, and about 30 cm of water above it', he says. 'So we really sampled the water/sediment interface.' In the clear pipes, he tells me, he could clearly see organisms swimming around, 'frantically saying, "What's going on?"'

With extreme care, the sediment is pushed out of the tube and sampled at 0.5 cm intervals, each representing about ten years of sedimentation, sometimes less. The known relationship between the chemical compositions of some algae relative to certain sea temperatures can, he says, be used to determine the incidence of those temperatures going back a very long way.

'I was able to tell, for example, that the sea surface temperature offshore Brisbane over the last 350 years has changed very little, whereas off Fremantle it has at times been both warmer and colder.' It has also led him to some intriguing conclusions regarding the current discussion over the behaviour of the East Australian Current. 'People are today saying that the East Australian Current is getting stronger, that it is bringing warmer waters to offshore Sydney and Tasmania. Well, we have evidence that this started well over one hundred years ago.' These findings prompt me to enquire further as to his stance on the topic of global warming generally. There is an uncharacteristically long pause before he answers.

'I am extremely careful here', he begins, 'but I will say the oceans are not warming uniformly. There are some hot spots such as south-east of Sydney, but west of Tasmania there's some cooling and we don't yet understand why. Some places are getting warmer, some are getting cooler or not changing at all. People think the oceans are getting 1° warmer everywhere. It's not entirely correct.' Patrick is confident the margin of error in his method is no more than one or two degrees. He is still working on the manuscript that will illustrate his research and is reluctant to pre-empt his findings, although he believes meteorologists 'ought to be more careful when stating that the oceans are warming up in regards to the question of sea temperature warming'.

More than anything Patrick has discovered about oceans, ancient rivers and sea temperature, it was the future he seemed most excited about, particularly the training of tomorrow's scientists. 'The *Southern Surveyor* was a good ship and everything worked well, but she was old', he says, 'with not much room for scientists and students. The *Marion Dufresne* can take a hundred people, which is why in France even undergraduates get to go to sea. We haven't been able to do that here, and that's why the *Investigator* will be such an important vessel for research as well as training.' Suddenly the word 'campaign' being used to describe Patrick's work seems appropriate.

Colin Woodroffe

Coastal geomorphologist

> *'I'm not a good sailor, but being Chief Scientist, you have to keep that as quiet as possible!'*

The array of images generously given to me by the scientists and crews who had worked on *Southern Surveyor* during her long and eventful life with the MNF was truly amazing, although sometimes I needed help deciphering what I was looking at. There were pictures of the CTD rosette, studies of the *Surveyor*'s powerful winches, drums of shiny new cable and arrays of enormous tools which looked like they could only be wielded by the hands of giants, and photos of other mystifying pieces of equipment with uses I could only guess.

Some brilliant sweeping aerial angles of *Surveyor* were also at my disposal: there she was, crashing through the swell of a Southern Ocean storm, gliding effortlessly across a glassy Arafura Sea, or departing from her pretty home port of Hobart for yet another discovery in marine science. There were the pictures of her times in dry dock, looking somehow vulnerable as holes were cut into her hull to allow important installations such as the swath mapper, its bulbous steel appendage hanging oddly beneath her. There was the herculean A-frame being welded to her back deck, allowing so much of the deep sea exploration she achieved to take place.

But among this mixed bunch of images were some which caught my attention immediately. One of those was not strictly of *Surveyor* at all, but a snap taken through one of her portholes – low and close to the water – of the stormy ocean outside, a ghastly grey seascape of wind-blasted white-caps and angry swells. Just looking at it one could sense the loneliness of being out there, in bad weather at the bottom of the world, a long way from home.

One series of images I came back to time and again, as they seemed to be of one of the strangest-looking places on Earth.

Just south of Lord Howe Island lies Ball's Pyramid, a gigantic blade of basalt as tall as the biggest skyscraper, jutting over 500 m into the sky like a terrible black sail of rock. With its impossibly pointed, threatening-looking spire, it seems to resemble no earthly place at all, rather the lair of some dark master in a *Lord of the Rings*-type

Looking through a porthole, on an expedition to the Northern Tonga Vents to investigate submarine hydrothermal plume activity and petrology in October and November 2004. Source: MNF.

fantasy. But Ball's Pyramid is very real, and *Southern Surveyor* visited it several times on important scientific research voyages. The pictures, after all, are there to prove it, although it took me quite a while to find someone who'd been along for the ride. Coastal geomorphologist Colin Woodroffe proved to be the man.

Despite (or perhaps because of) being originally from England, Colin's early research centred on the tropics. Reefs, estuaries and mangroves were all his area of interest, with a PhD on the mangrove swamps of the West Indies and Cayman Islands. For the last twenty-five years, however, he has been part of the University of Wollongong's School of Earth and Environmental Sciences, with a few years in Auckland and Darwin squeezed in there somewhere as well.

As he begins to reveal some of the remarkable facts about Ball's Pyramid and nearby Lord Howe Island, his enthusiasm for these remarkable places becomes infectious. '*Cool!*' I find myself exclaiming. 'I know, it's pretty amazing, isn't it?' he replies, both of us sounding like schoolboys exploring a mysterious cave.

Even without knowing a skerrick of its history, Ball's Pyramid is still impressive, but the fact that it is the core of an ancient, all-but-eroded volcano makes it to me, and to Colin, even more amazing. 'It's the classic example of a volcano that has been truncated by wave erosion', he tells me, 'so all that remains is this spectacular monolith of basalt which rises up out of ocean about 550 m.' Part of it, he says, is the remains of the inner magma chamber, as indicated by some vents which have been discovered

The sun sets behind Ball's Pyramid, the remnant of a volcano near Lord Howe Island. At 1100 m, it is the tallest volcano stack in the world. Source: MNF/Kim Picard.

deep in its mass. The whole thing appears to sit up on a shallow underwater platform in just 30–50 m of water, which extends several kilometres around it on all sides. This is the remains of its own ancient footprint, the last remnants of its once gigantic dimensions before being slowly eroded by six million years of wave action.

To demonstrate the relative scale of these islands and the platform on which they sit, Colin shows me one of the larger swath map images – taken on *Surveyor* – of the entire region, and it is extraordinary. One can clearly see the base of what was an almost unimaginably large volcanic mountain, sliced off at the base like a knife, with only the relative pinpricks of Lord Howe Island and Ball's Pyramid poking, like matchsticks, above the surface of the sea. What the original mountain would have looked like six million years ago almost defies imagination. It is equally dramatic at the edge of the shelf, where it plunges away over a series of rugged, almost vertical, ravines to depths of several kilometres off the continental shelf.

As dramatic as the forms of Ball's Pyramid and the ruggedly volcanic Lord Howe Island are, it is their position – right on the southern threshold of those parts of the ocean where coral reefs form – which intrigues Colin even more. Over millions of years, he says, the area has been gradually moving north into the warmer, reef-bearing oceans, providing scientists with a unique opportunity 'to look at volcanoes moving into these reef-forming seas'.

'Waves are very effective at eroding volcanoes, even truncating them completely, as with Ball's Pyramid.' Reefs, on the other hand, make an excellent protective barrier. He cites the examples of Hawaii and the Society Islands where coral atolls formed around ancient volcanoes, allowing them to subside and disappear naturally over the millennia 'in what we call a Darwinian sequence'.

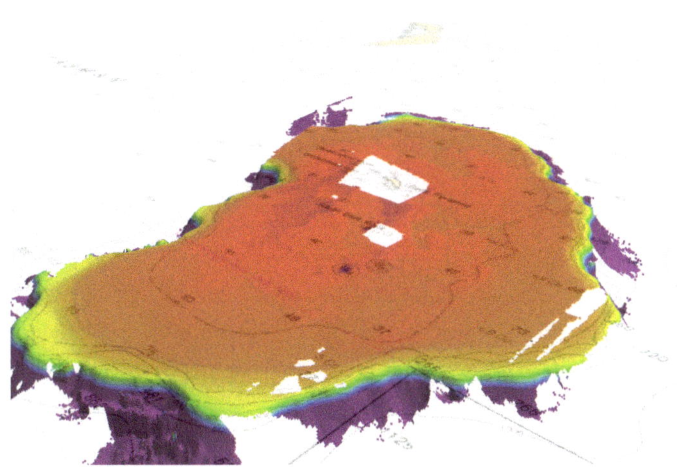

Multi-beam sonar image of the region around Ball's Pyramid (upper grey square area), showing the size of the ancient volcano. Source: MNF.

Other areas around Lord Howe Island also interested Colin. Elizabeth and Middleton reefs, part of a relatively shallow plateau called the Lord Howe Rise, lie over 100 km north of Lord Howe Island itself. They have done very little over the past few hundred years but get in the way of several unlucky ships, but may provide some clues to the volcanic history of the wider area. Although far from the tropics, they bear some intriguing similarities to many of those more northerly atolls and may, says Colin, have once similarly protected long-vanished volcanoes. However, whether these were gradually eaten away by wave action, as with Ball's Pyramid, or have simply subsided slowly was one of the questions Colin attempted to find answers to on his two voyages with *Southern Surveyor.*

One of the most surprising results of Colin's voyages was of the discovery of the make-up of the shelf itself. 'We were expecting an eroded volcanic rock platform', he says, 'but found from the swath mapping on the *Surveyor* these prolific carbonate environments – substantial fossilised coral reefs about 9000 years old. This coincides with the time that sea levels would have been about 30 m lower than today.'

During the Ice Age, he says, this exposed reef would have ringed the entire island, considerably reducing the power of the waves. 'Ten thousand years ago, Lord Howe Island would have looked more like Tahiti. It was very exciting to find there had been reefs around Ball's Pyramid, as there has always been a presumption that during the Ice Age, the extent to which reefs such as this could grow would have been considerably limited.' Colin even found small colonies of living coral, making them the most southerly examples of living coral to be found in the Pacific.

The first of Colin's two *Surveyor* voyages focused on Lord Howe Island and was undertaken a few months before the terrible 2008 Sydney–Hobart yacht race disaster, when the fleet was decimated by deadly storms. Colin describes the weather he experienced as being similar. 'Horrendous', he says, '6 m swells with the ship hove-to in the middle of the Tasman Sea, and that's very common for Lord Howe Island.' Being a place where it's difficult to find any sort of calm weather at any time of the year, even at the height of summer when he ventured there, Colin regards himself as lucky to have been able to drill the six cores that they did. 'The captain needed to be convinced that the weather was good enough to be able to keep the ship on station, and deploy the rig', he says.

The voyage in February 2013, which Colin led as Chief Scientist, was one of *Southern Surveyor*'s last. It concentrated on Ball's Pyramid and was blessed with a period of exceptional good weather before reverting to type and becoming 'as bad as it gets. We had five or six very rare, very good days, then it all closed in.' Their second week was spent more or less caught on the western side of Lord Howe Island where, says Colin, 'all we could do was a little bit of grab sampling of the shelf, but nothing else'.

Colin confesses to feeling the effects of the weather, but says the responsibility of being Chief Scientist on the voyage to Ball's Pyramid assuaged some of the discomfort, or at least encouraged him to conceal it. 'I'm not a good sailor', he says, 'but being Chief Scientist, you have to keep that as quiet as possible! Anyway, after a few days of being buffeted around you find your sea legs. I don't think I shamed myself, there were certainly people onboard a lot worse than what I was!'

Discomfort notwithstanding, the work Colin and his team – including several rather hardy students – did manage to carry out was extensive, utilising both the rock drill which sits on the bottom with a waterproof motor and is operated from the deck, and the three-legged vibro-corer for the soft sediment, where the evidence of the once extensive coral reef became evident. 'We found very little of the volcanic material around the Pyramid', he says. 'Instead it was mainly the white carbonate sediment from the corals and coraline algae which can tolerate cooler temperatures, which is why we're finding them here beyond where corals are more prominent.' Despite this, Colin still smarts at that 'lost' second week of his second voyage. 'Often, life on the ocean is a challenge', he says with that familiar tone of resignation I've heard from many marine scientists whose best-laid plans were foiled by the weather gods. 'You just can't always do what you set out to do.'

Swath mapping was also carried out on the shelf but, at depths of only 50 m or so, this presented its own problems. 'Because of the shallow water', says Colin, 'the width of the swath was quite narrow, and so we had to go up and down, back and forth many times to get a complete picture.' It also happened to be uncomfortably near the spot where, in 2002, the British warship HMS *Nottingham* ran aground on a place called Wolf Rock, tearing a hole from bow to bridge, nearly sinking one of Her Majesty's destroyers, ending several promising careers and causing a public-relations headache for the Royal Navy. The *Surveyor*'s captain was understandably nervous.

The vibro-corer is a long steel tube that vibrates to push it up to 6 m into the seafloor. It is used on sandy sediments where other corers are not effective. The long and short sediment corer can take samples of other sediments up to 24 m in depth. Source: MNF/Brendan Brooke.

'We came pretty close to an outcrop called South-east Rock that sticks out of the water there and the captain eyed it nervously', says Colin, 'but as we collected the data, he gradually became more confident he wasn't going to hit something. He really didn't want to leave the ship grounded on a rock as a museum piece on one of her last voyages!'

After six million years of erosion, the mighty Ball's Pyramid still stands, towering majestically over the sea on its own shelf, gazing out into the Pacific, as it has done for aeons. Will it, I ask Colin, one day succumb to the power of the waves and just topple over? 'That's a fascinating question', he says. 'In six million years, that much, we know, has eroded.' But as the Pyramid shrinks and the shelf on which it stands gradually enlarges, the wave action becomes more and more dissipated and the rate of erosion slows. It may fall, Colin suggests – I'm sure it will be spectacular, but I won't be putting money on it happening in my, or even my great-grandchildren's, lifetime.

On balance, Colin says he was 'absolutely delighted' with the Lord Howe Island and Ball's Pyramid voyages and would like to think there will be further opportunities to continue its exploration in the future. 'We certainly haven't answered all the questions there.'

Andrew Heap

Marine geologist

'If they find something, and I'm pretty sure they will, there'll be billions of dollars of revenue coming back into Australia.'

Looking for oil and surveying potential marine parks appeared to me to be two quite unrelated activities, but not after talking to marine geologist Andrew Heap. Working with the Australian Government's national geoscience agency, Geoscience Australia, has provided Andrew with a varied career to say the least, and on a cool sunny Canberra afternoon he quickly runs off a list of things that have been keeping him busy lately, many achieved onboard *Southern Surveyor*. 'We do geoscience studies that underpin oil and gas exploration; environmental work that helps establish marine parks; groundwater studies; work on natural hazards such as landslides and volcanoes; looking for undersea areas to store carbon dioxide; oh and we also look after our maritime boundaries so we know where our jurisdiction lies. Oh, and we discovered two new islands', he adds intriguingly.

It appears that these reef islands, situated a couple of hundred kilometres off the Western Australian coast and measuring a not insubstantial 15 km long and about half that wide, were always known but not included in the state's jurisdiction. The states, by tradition, control offshore islands and their surrounding areas to a distance of around 3 km but, except for a few colonies of seabirds, these nameless reefs had been all but forgotten. Only with the discovery of oil and gas has the ownership of these islands become important, as they lie conveniently close to a recently discovered undersea gas field.

Surveyor's work in determining that these islands were indeed the property of Western Australia resulted in a slight rewriting of that state's border, and the doubtless undying gratitude of the WA government which will in time reap rewards from the ensuing royalties. (Andrew's still waiting for their offer of naming rights, but he suspects the wait will be a long one.)

Andrew is most complimentary towards the good lady *Southern Surveyor*, having undertaken twelve very different voyages on her. 'A very capable vessel' is how he

'A very capable vessel.' Source: MNF/Richard Arculus.

confidently sums her up. 'For her age and her size, we got a lot out of her. It was stable, and it could go to places other vessels couldn't. When I compare it with some other overseas ships I've worked on, we really did manage to do a great deal on her.'

One of those achievements was a two-year multi-beam mapping survey of Australia's offshore regions undertaken for the federal Department of the Environment, to gather information for the establishment of Australia's marine reserves under the Howard Government's Oceans Policy in the early 2000s. The *Surveyor* was used to do a good deal of this work: mapping the seabed, taking core samples, video samples, ocean water chemistry samples – all to characterise the seabed and outline the different underwater habitat types. Before this more specifically scientific approach, Andrew tells me, decisions governing the borders of marine parks evolved more or less along the lines of 'That looks like a nice coral cay with some fish around it.' 'But beyond what you could see', he says, 'there was very little appreciation.'

'We mapped the continental shelf, the slope and also the deeper parts: the canyons, the seamounts, even the flat and boring abyssal plains', he says. 'It's all about outlining the different underwater habitat types because these have a bearing on the plants and animals that live there. On land, it's easy, but when it's covered with water and permanently dark and below 100 m, it's very difficult.' And unless we know what's there, he says, we don't know what's there to manage.

Bringing up some of these swath images on his laptop, Andrew moves his mouse over the myriad colours denoting various depths and gradients of the seafloor, then

AUSTRALIA'S MARINE PARKS AND RESERVES

Marine protected areas, marine parks or marine reserves are areas of the ocean around the world protected by national, state or local governments.

Australia's marine Exclusive Economic Zone covers over 13.86 million km^2, and is the third largest in the world. Commonwealth marine reserves are located in waters that generally extend from three nautical miles off the coast to the outer limit of Australia's EEZ (200 nautical miles), with marine protected areas closer to shore the responsibility of the states or Northern Territory.

Marine reserves are areas where the ocean is protected from threats, to help maintain healthy ecosystems. They are not about closing fisheries or preventing exploration, but about preserving ocean health and managing its use. The UN Convention on Biological Diversity, of which Australia is a party, urges the conservation of 10% of a country's coastal and marine areas by 2020. Australia's Commonwealth marine reserves account for more than a third of Australia's EEZ.

Marine reserves help protect the health of Australia's oceans. They protect areas of high conservation value, fulfil Australia's international commitments for biodiversity protection and provide a science-based framework for management of the oceans.

Notable examples of Australia's more than two dozen marine parks or reserves include Ningaloo Reef, Lord Howe Island, Heard Island and McDonald Islands, the Great Australian Bight, Shark Bay and the Huon (Tasmanian Seamounts) Marine Reserve. The Coral Sea and adjacent Great Barrier Reef Marine Protected Area together create the largest marine reserve in the world. Many of these were visited by RV *Southern Surveyor.*

Snorkellers keep up with a whale shark as it swims at Ningaloo Reef, Western Australia. Source: Julie Edgley (CC BY-SA 2.0).

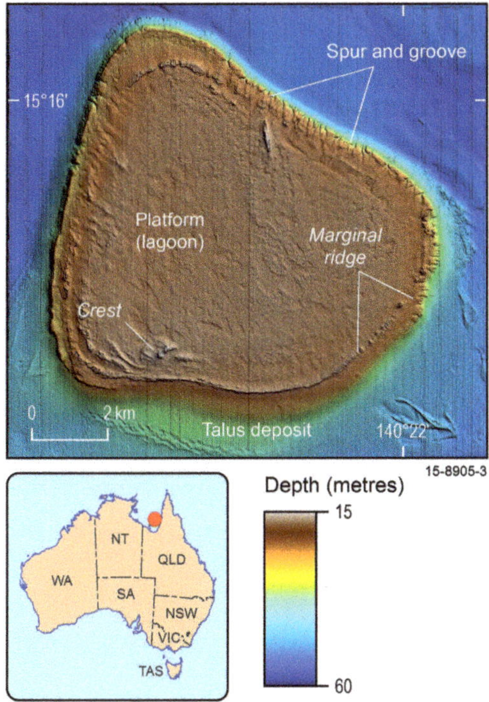

Multi-beam sonar image of a 72.5 km² previously undescribed submerged carbonate reef in the southern Gulf of Carpentaria mapped by Geoscience Australia during a *Southern Surveyor* voyage in March 2005. Source: Geoscience Australia.

pauses. 'Oh, and these are some submerged coral reefs we discovered in the Gulf of Carpentaria', found when they weren't even looking for them.

'It was quite interesting actually', Andrew begins. 'We were out doing sediment cores up and down the length of the Gulf to learn about its history over the last 20 000 or so years. On the chart, there were a couple of high points we decided to explore.' The Gulf of Carpentaria, although large, is not particularly deep, with an average depth of about 70 m. Any raised areas stand out. The fishermen knew that those places were reliable sources of mackerel and other species, but had no idea why. As the *Surveyor* began to map and the swath images started to emerge, one small tantalising strip at a time, Andrew and his colleagues were taken by surprise. 'That looks like a coral reef', they remarked, and so decided to map the whole thing. Sure enough, an undiscovered coral reef is what it turned out to be.

Video footage provided visual evidence of these previously unknown reefs teeming with life just 20 m below the surface of the Gulf of Carpentaria. A total of seven were discovered on this voyage. 'There are more, but we just didn't get time to map them', he says. But when they took some core samples, there were more surprises. 'When we dated this coral, we found that it was old, older than the Great Barrier Reef', although nothing like the scale of that famous landmark. The coral was rich in parts but patchy, interrupted with large sandy areas. 'We think', Andrew tells me, 'that as the sea levels rose at the end of the last great Ice Age, the Gulf of Carpentaria filled up from the Arafura Sea from the west not the east, so predating the Barrier Reef.'

I wondered, though, if these corals were indeed older why were they less established and why had they not even reached the surface, remaining instead at 20–30 m below? The answer, Andrew suspects, lies in the Gulf's low nutrient levels, but this remains to be determined by future voyages. However, thanks to the work of Andrew and his colleagues, these reefs have been incorporated into Australia's marine park network.

It was on a subsequent voyage exploring the Arafura Sea that another discovery was made, this time in the form of what Andrew describes as the 'great carbonate banks'. 'Ginormous' is the word he uses. 'We found the whole area filled with them – made of shells, coral, sponges, all sorts of things, about 30–40 m high, and no one knew about them.' Banna Bank is a huge underwater natural feature beginning just 20 m below the surface and sloping away to the floor of the Arafura Sea. 'We mapped it all out – the habitats, the plants, the fish, particularly the schools of swordfish we found there', he says.

The genesis of these banks, I learned, lies in the fact that Australia is a somewhat 'leaky' place in terms of its seafloor, particularly in the north-west, a phenomenon which people such as Andrew on *Southern Surveyor* utilise to search out places where petroleum might be found. 'Signatures of hydrocarbon seepage' is what he calls it, chemical indicators of potential petroleum reserves which sit in rock formations deep in the sediment and occasionally leak their way up to the seabed and into the water. His colleague Neville Exon had told me about such occurrences in southern Australian waters around Tasmania, explaining how they can reach sizes of 0.5 m or so. 'What we were trying to do was target those seeps so we could then determine the hydrocarbons, and look to see where there might be a working petroleum system', is how Andrew describes it. Put simply, they were looking for signs of oil.

They're not always successful. On this occasion, after samples of water and sediment were taken, little other than background evidence of petroleum was found but Andrew was more than happy to have discovered these new habitats nonetheless. Their presence is related to hydrocarbon seepage. 'The nutrients that come with it are used by the carbonate organisms to grow these banks. They use them for food.' It seems to me a curious symbiosis – leaking, oil-based sea chemicals on the one hand and teeming marine environments on the other, but it illustrates another aspect of *Southern Surveyor*'s work.

Perhaps one the most interesting discoveries, at least in terms of its potential, was made at the opposite end of the continent, in the Great Australian Bight. It involved the somewhat surreal story of a river which, says Andrew, was once as big as the Amazon and flowed through what is now the parched outback. He pores over a map of the Bight and points to a series of lines and coloured patches. 'It's called the Ceduna Terrace', he says. 'A huge river delta.' At least it was, a very long time ago.

In the Permian, Carboniferous and Jurassic periods, 350–150 million years ago, a far more tropical Australia produced a mighty river which flowed through what is now the barren South Australian desert. It deposited in its wake an enormous pile of sediment 8–9 km thick. 'We've done quite a bit of work on those sediments', says Andrew, 'and identified where there might be some very nice petroleum systems.' It was a find made off the back of an earlier, less successful investigation by another vessel in the early 2000s. 'They went out there and put down a hole but, because of the weather, missed their main target and declared it was all too hard and that there was nothing out there.'

Surveyor, however, was not deterred and had an advantage over some other ships, stemming from the legacy of its previous life. 'Being an old fishing trawler', says Andrew, 'she has these wonderful winches used for hauling nets, so we used them for rock dredging.' It's not the most subtle way to explore the seabed, scraping and dragging a steel cage along the bottom to collect rocks, but it provided some remarkable results. 'The skippers were ex-fishing captains who had all trawled, and were comfortable in dragging a great heavy thing along the seabed.' Other vessels Andrew has travelled on were more reluctant to allow themselves to be directly connected to the seabed by a very heavy piece of steel and a cable!

Within this delta system is a series of very deep canyons cut by ocean currents and the movement of the Earth. 'A little like the Grand Canyon but under the sea. We targeted some specific depths in these canyons and brought up some very dark and organic oil-prone rocks.' Geochemical analysis revealed enormous indications of oil and, as Andrew rather enthusiastically puts it, it has the potential to 'change the world' in terms of Australian oil exploration. 'If they find something, and I'm pretty sure they will', says Andrew, 'there'll be billions of dollars of revenue coming back into Australia.' Not a bad return for an investment of around $6 million.

Of his dozen voyages on *Southern Surveyor*, Andrew says that no two were the same. 'You'd go out with all these preconceived ideas about what you would do and what was going to happen, but in a couple of days, they're just blown out of the water, literally.' He cites his two trips north, to the Gulf of Carpentaria and the Arafura Sea, as his most memorable, but it is his work on the discovery of the Gulf's submerged reefs of which he is most proud. 'That's the one that really stands out for me. We found something completely new, we answered a global story about reef development, and helped design a marine park', he says. 'Oh, and it was beautiful weather.'

The fact that Andrew has contributed to the establishment of some of Australia's marine protected areas will, I sense, remain one of his most important achievements as a scientist. As he proudly but plainly states, 'As a result of a lot of the work that we did onboard *Southern Surveyor,* Australia still has the largest marine park network in the world.'

He is greatly appreciative of the special breed that made up *Surveyor*'s crew, who 'weren't afraid to do what you asked them to do. You've got pointy-headed scientists asking all sorts of things and most of the time the crew would oblige, in fact they would take it almost to the limit to get what you needed to do the science.' As an example he cites a typical attitude of skipper Ian Taylor. Scientists are always wanting to get close to things, particularly reefs, but these pose problems for curious ships. 'Ian was terrific', says Andrew. 'A couple of times I'd see a reef and ask him, "How close can you get to that reef?" And he'd turn to me and just say with a grin, "How close do you want to get?"'

Steven Micklethwaite

Geologist/geophysicist

> *'It all goes to explain why the Coral Sea is the way it is, and why Australia formed in the way it did.'*

Southern Surveyor was responsible for a great number of discoveries: massive, previously unsuspected seamounts, undersea volcanoes, shipwrecks, scientific theorems regarding currents, ocean warming, marine debris and biomass and, as Andrew Heap told me, the 'discovery' of two islands off Western Australia. *Surveyor* has also been credited with 'undiscovering' an island, a feat which, when it happened, grabbed headlines around the world.

The intriguing story of the undiscovery of Sandy Island remains a unique chapter in *Southern Surveyor*'s history although, like many of the world's great discoveries (indeed, undiscoveries), it was all an accident.

Structural geologist Steven Micklethwaite spent just one voyage aboard the *Surveyor* but it's one he'll remember, and not just for the quirky story of Sandy Island. Steve's field is plate tectonics and the movement of continents. From his study of the formation of the islands of Indonesia, his interest progressed to the tectonic history of the Coral Sea which, it transpires, is a most unusual part of the world.

I've always rather liked talking to geologists, mainly because they never fail to present me with at least one completely staggering fact. Steve gave me three. We talked a while about the Earth and its formation, about how the plates move at different rates and in different directions, like gigantic vessels tracing indescribably slow-moving paths across the face of the globe. Then he tells me that the very earliest rocks – formed during the Earth's first billion years – are chemically different from those that came later, due to the Earth being so much hotter back then. Rocks such as the extremely rare komatiite, for example. And there is a place in Western Australia where Steve works (at the University of Western Australia) called Jack Hills where the oldest rocks on the planet have been found, dating back a staggering 4.5 billion years – to the very formation of the Earth itself.

He also told me how the Earth's magnetic field can be locally affected by the rocks in individual areas, and that by mathematically filtering the signal a map can be generated showing their different magnetic characteristics. This can be done on land from an aeroplane, or at sea by towing a torpedo-shaped instrument called a magnetometer behind a vessel such as the *Southern Surveyor.*

'A weird ocean', is how Steve describes the Coral Sea. 'When you look at the shape of it, the Coral Sea has some really unusual features.' This, essentially, was the reason why, in 2011, a group of young scientists, led by University of Sydney geologist Maria Seton, applied to the Marine National Facility Steering Committee for the use of *Southern Surveyor,* and were accepted.

'The objective was to go and collect rock samples, to study the bathymetry and other high-resolution data from across a good portion of the Coral Sea to work out how it formed', says Steve. Global bathymetry images of the Coral Sea reveal certain oddities which, he says, are not seen in many places elsewhere. 'Ridges, big flat expanses and plateaus emerging out of nowhere, and it's very hard to understand how all these different features formed.'

Not only is the exploration of the Coral Sea's history of scientific interest, says Steve, it tells something fundamental about how Australia evolved – 'Why we are the way we are', as he puts it. It's also a story of national significance – identifying the ocean floor and which parts of it might belong to the Australian continent has the potential to influence claims of fishing resources and future energy reserves. Quite a hefty responsibility to put in the hands of a particularly youthful bunch of scientists.

'The team was very unusual in that we were all very young', says Steve. 'Normally, there's a more experienced member leading it up, but because we had such an interesting idea, and the *Surveyor* was coming to the end of its life with the MNF, I think the funding bodies thought they'd give us a chance.' They weren't disappointed.

Departing from Brisbane, the group spent 23 days at sea onboard *Surveyor.* Seventeen sites over the Coral Sea's vast expanse were mapped out, with twelve dredges eventually made. It was a well-oiled routine. Pairs of scientists working in shifts oversaw the deployment and towing of the magnetometer. When a designated position was reached, the area was mapped in detail then a dredge was sent down to collect rock samples from the bottom.

Only after the *Surveyor*'s crew had secured the heavy steel dredge safely onto the back deck were Steve and the other young scientists permitted to pore over this catch, ripped suddenly from the sea bottom several kilometres below. Sifting through the sludge and stones and other detritus, they sought eagerly for something which could unlock the many secrets of the ocean's geological and tectonic history, particularly rocks displaying signs of volcanic origin. These arrive from the deep, Steve tells me, covered in 'a very fine mud' which has to be washed off, then a thick layer of manganese oxide which has, over the aeons, precipitated from the ocean floor and formed on the rocks like a hard rusty skin.

Geologist Damien Ryan studying a freshly dredged-up rock. Source: MNF/ Cameron Mitchell, Geoscience Australia.

'The chemistry of lavas can tell us whether they came from a subduction zone, a mid-ocean ridge or the middle of a continent', says Steve. Hence, once the rocks were onboard and selected, there were many hours of washing, cleaning and cutting in the small confines of *Surveyor*'s laboratory.

The scientists are also, of course, eager to determine the rocks' ages, using a technique called argon-argon geochronology, a radioisotope technique used to measure concentrations of different isotopes of the element argon.

Also of interest were the limestones, which were handed to the University of Tasmania's Pat Quilty, who, by examining the fossils, could determine the stone's age. 'Gradually we began to build up a picture of the rocks – their chemistry and their age – and start to determine how the Coral Sea was formed and what processes went to make those very funny shapes', says Steve.

With patient and detailed detective work, Steve and the team have been able to build up, for the first time, the beginnings of a picture of the history of this 'weird ocean'. 'The seafloor has been going up and down with time', he says, giving me an impression of some vast primordial trampoline. 'Lavas erupted and were pushed to near the Earth's surface where limestone formed on them, making them sink again.'

It's also the history of volatile volcanic activity. 'It's had at least two chains of volcanic underwater ocean ridges, one of them probably started by a giant volcanic plume, a hot-bodied piece of mantle rising underneath the Earth and causing unusual melts and magmas, which then migrated away', he says. 'It all goes to explain why the Coral Sea is the way it is, and why Australia formed in the way it did.'

But to the most peculiar story of Sandy Island. With seventeen dredge sites planned to be visited over 23 days across an area the size of the Coral Sea, *Surveyor* did a great deal of sailing and her routes had to be carefully planned. The process was simple enough: a line was drawn between two places on the map, the distance roughly calculated, then shown to the captain whose job it was to get them there. 'Actually we use Google Earth', Steve says, to my amusement.

On one occasion, when planning a route out to the west of New Caledonia, Steve noticed that it took them past an unusual-looking feature with the startlingly imaginative name of Sandy Island. On the map, Steve thought it looked like a giant snake eye. It was not insubstantial, measuring 5 km wide and 25 km long, about the size of Manhattan. Strangely, Google Earth provided no visual details about Sandy Island. There was no satellite photo nor any other scrap of information which might indicate something about its topography. It was simply black, like a long slit in the middle of the ocean.

There were a couple of other things which didn't seem to add up either. 'Google Earth gives you the shape of the ocean floor', he says, 'and in this part of the Pacific it was 1400 m deep. Then, all of a sudden it rose, straight up out of the water to this Sandy Island. It was like it just jolted out of the Earth like a giant pedestal.' Strange as it appeared, the island's existence was nonetheless confirmed by several other sources, weather maps and atlases used by scientists and mariners. It even appeared to be confirmed by the highly respected international World Vector Shoreline database. 'The only place it wasn't', says Steve, 'was on the French and Australian navies' navigational charts, used by the *Surveyor*.'

Presenting their conundrum to the *Surveyor*'s captain, they decided to sail a route which would take them straight through the middle of the island, if it existed. 'The captain was extremely nervous. It didn't exist on his chart, but on his weather map, there it was', says Steve. Safety procedures were put in place, the bridge was filled with extra pairs of eyes keeping a lookout through binoculars, and an extremely cautious approach to Sandy Island found that in fact nothing was there. 'The sea never got shallower than 1300 m', says Steve. The *Surveyor* sailed right through the middle of Sandy Island without running aground.

The atmosphere onboard the *Surveyor* upon discovering the non-existence of the island was ebullient. It is not every day one gets to change the map of the world, even if it is to remove rather than add something to it. But who is to say that proving something not to exist is any less valid, scientifically, than proving it does? Steve concurred with my theory completely.

Satellite image showing the location, near New Caledonia, where Sandy Island was previously recorded to occur. Source: MNF.

Releasing this tidbit of world geography to the media, Steve and the team were amazed to see that what they regarded simply as a funny story raced around the world like lightning. 'It went viral', he says. 'I was interviewed on TV by Channel Four in the UK via Skype, the ABC – it even made news items in countries like Malaysia, Saudi Arabia and Belgium. Even before we landed back in Brisbane it was a media story.' In 2012, the story 'Where did it go? Scientists "undiscover" Pacific Island' became the single most-read story on the website of the *Sydney Morning Herald*, beating an impressive line-up including the US elections, Hurricane Sandy and the release of the Apple iPhone 5.

How did a non-existent 25 km island which someone had even gone to the trouble of putting a name to, get stamped into some of the world's most respected cartographical publications in the first place?

'That's a story of citizen science', says Steve. Within about a week, amateur sleuths and enthusiasts around the globe had researched and pieced together the puzzle. It seems that around 1906 a whaling vessel had reported an island in this location, but only as a possible sighting. 'The island was mapped, but never confirmed', says Steve. It did, however, disseminate into worldwide community datasets which, upon the advent of the computer age, were incorporated into behemoths such as Google Maps, which in turn propagated the false information all over again.

'Google Earth actually came to us and within twenty-four hours it was removed from Google Maps', says Steve. 'We believe that they had actually been told previously by the body that oversees Australian navigational charts, but that message seems to have gotten lost in the ether.' We go on to discuss the prospect of the outraged French, demanding compensation for their imaginary Pacific outpost!

Most of the information gleaned about the history of the Coral Sea, Steve tells me, has yet to be published, but he is confident it will add substantially to our body of knowledge concerning this important and unusual area of ocean to Australia's north-east. 'You probably won't see the impact of our findings for a couple of years but it will be very interesting when it comes out.' More important than this, Steve has a final message for me. 'If you do one thing in your book', he says, 'praise the crew. They were amazing.'

Even the food was exemplary, he says. 'We were away for twenty-three days and on the very last day they were still serving things like barramundi and prawns.' From the deck crew to the chefs to the people who kept the engines running, says Steve, everything worked particularly well. 'They were all professional and friendly', he says. 'And if they weren't friendly, they stayed out of your way.'

Richard Arculus

Igneous petrologist

> *'The deepest trenches, highest mountains, biggest earthquakes, most explosive volcanoes are all associated with these places. We're discovering things all the time.'*

Professor Richard Arculus' reputation precedes him like the cutting bow of a ship travelling at full steam across a wide ocean. Not only did every single person I spoke to in connection with *Southern Surveyor* seem to know this remarkable man and have not a bad word to say about him, they seemed unable, in fact, to speak of him in anything other than glowing, almost ebullient terms. 'Brilliant', 'amazing', 'inspirational', 'extraordinary mind' the accolades went. But just try tracking down the Professor of Geology in the Research School of Earth Sciences at the Australian National University. 'Oh no, he's in St Kitts', someone told me. Then I heard he was speaking in the US, or that he'd most likely be found in Iceland or perhaps the Caribbean, or on a ship in some obscure part of the Pacific Rim where he has built a worldwide reputation in the study of undersea volcanoes.

I met Richard in the foyer of his department. He shook my hand warmly while dealing with a phalanx of adoring students, some with exotic accents, all of whom were anxious – in hushed tones – to speak with him, and to whom he in turn appeared to show equal consideration. He led me into his office, a cramped space bursting with racks of rock samples, maps and documents of all kinds, and a desk dominated by what is I think the largest computer screen I have ever seen.

After we sat down and he made me a cup of tea – checking carefully as to how I liked it – I decided to hurl myself into the deep end with a prepared, though somewhat grandiose question. 'Does the Earth hold on to all its history?' If anyone could answer it, I thought, it must surely be Richard Arculus, as he has spent so much of his life seeking out the Earth's secrets. There is a rich and palpable pause as he looks at me without the faintest sign of expression. A quiet 'Mostly … mostly …' then like a powerful steam train gradually pulling away from the station, he launches into a brilliant and wide-ranging exposition. 'The bonus – from the point of view of an

Earth-bound geologist working with an active Earth's surface – is that it's very interesting. The drawback, of course, is that things get reworked over and over again.' At this point he suddenly bounds up and shuts his office door ('Otherwise we'll keep getting interrupted by people wanting to talk to me'). 'Geological events are overlaid by subsequent geological events. Unravelling them almost becomes a police problem, a CSI-type of thing.'

'Here, I'll just draw you an abundance plot of the elements in the Sun. Just for your general knowledge', he says, diving back into his exposition. 'We'll make this a logarithmic scale going from hydrogen to uranium …' It's not every day someone makes you such an offer and he is genuinely keen for me to understand, but I'm struggling to keep up.

What are the origins of such a lively and active mind, I wonder? I ask Richard about his upbringing. Many men of his generation would cite their father's influence but instead he speaks of his mother, a truly remarkable-sounding woman. 'She was part of the British Raj', he tells me, born, like Richard himself, in India. Sent back to England for her education, she was offered a job at Cambridge, found herself in Germany in 1939 and spent the war with Alan Turing and the legendary codebreakers of Bletchley Park. 'When you go there', says Richard, 'there's a photo of the people from Hut 3 on VE day. You can see my mother there on the right. She was still subject to the Official Secrets Act and couldn't even talk about it until about 1970. My father was a Cambridge-trained mathematician and became a major in the artillery in India', he adds. Meeting in the Far East, Richard's parents decided that they both liked India, and so decided to stay.

Like his mother, Richard was sent back to England to complete his education, but in the grey grime of post-war Britain he longed for the exotic expat life he'd left behind, which may explain how he eventually found a way to complete his PhD on the island of Grenada in the Caribbean.

Richard intended to work in industry, but 'a series of conveyor belts has taken me all my life through academia'. He can number himself among the alumni of such institutions as the Carnegie Institute of Washington, where he completed a postdoctoral thesis, the universities of Michigan and Durham, Rice University in Houston and of course the ANU where, on the rare occasions he is not at sea, he can be found in the small Aladdin's cave of a study in which I sit with him today. One of the drawcards which brought him to a permanent home in Australia was the opportunity to use the Marine National Facility. 'I realised that I could come back here and start to apply to use this asset, and go to sea. And it's been superb, Michael, just superb. It's been a major part of my life over the past fifteen years.'

'Almost all of what I've done', he tells me, 'not just with *Southern Surveyor* but previously with the *Franklin*, the New Zealand research vessel *Tangaroa*, as well as various American vessels, has been in the south-west Pacific looking at subduction zone systems, island arcs, back arcs, trenches.'

An isolated volcano breaking the water's surface at the southern end of the New Hebrides Arc. Source: Richard Arculus.

He explains that he is not a physical volcanologist as such, but an igneous petrologist, interested in the chemical make-up of rocks, how they form and the temperatures at which they do so. 'In other words', says Richard, 'I'm interested in how the Earth works, because believe me, we don't understand it very well.'

Starting out working on some of the great above-ground or 'subaerial' volcanoes such as Aso, Mt Fuji and those in the Pacific Ring of Fire, Richard in 2000 had a revelatory experience on his first MNF voyage with the *Franklin* to some of the islands in the Solomons. 'I just thought, "Wow, this is really where I should spend my time."' Switching to submarine work was, he says, transformative.

'The trouble with volcanoes that are in the air is that it's actually hazardous to go near them', he says. 'For example, trying to sample the gases coming out of a vent is incredibly dangerous. It's so much safer to grab them at 1500 m down in the water column, because the water acts like a big dampener.' 'Safe' can be a relative term, and one of Richard's earlier voyages – with the *Franklin* – is worth mentioning to illustrate the potential hazards of his job.

'We'd always wanted to take a look at Mount Kavachi in the New Georgia group in the Solomons', he says. The night before the ship arrived at the volcano site, the captain, Neil Cheshire, asked Richard what the chances were that it would be erupting. 'Minimal', Richard confidently replied. 'Only happens, maybe, once every ten years, shouldn't be any trouble.' The skipper nevertheless announced that he'd prefer to arrive in daylight in order to inspect the sea surface – just in case. 'And with good reason', says Richard. 'Some years before, a Japanese research vessel had been completely swallowed up inside an enormous volcanic gas bubble which broke the surface of the water underneath it. The entire ship simply disappeared.'

Due to arrive at Kavachi at first light, an excited Richard went to bed at around 2am. A little later there was a knock on his cabin door. 'Richard, you're wanted on the bridge', said the voice of the bosun. Still in darkness, with only the eerie red illumination lights designed to enhance the crew's night vision, Richard arrived on the bridge. 'Just look in the radar', said the skipper casually. Richard peered into the black rubber viewing cone but could see nothing unfamiliar. 'What do you want me to look at?' he asked the skipper. 'Just keep looking', was the reply.

Suddenly the screen was filled, says Richard, with 'all sorts of sparkly stuff'. Quickly glancing up at the windows of the bridge, he saw a fireworks show with lines of brilliant red was erupting out of the water off *Franklin*'s bow. '*That's a volcano*!' Richard exclaimed. 'Yes, it's Kavachi', said the skipper. 'It's right where we're headed!' Kavachi was erupting, and Richard estimated its active vent to be just *one* solitary metre below the surface. 'We could have hit it – klunk!' he says. This enormous natural structure rising 1500 m from the seabed was on the verge of breaking the surface. 'It was just beautiful', says Richard, 'but he was quite right not to trust me about it erupting only every ten years!'

Richard's work on *Southern Surveyor* involved the discovery of an array of submarine volcanic structures with curious terms such as 'young lavas', 'black smokers', 'hot spring systems' and 'cones', 'all that sort of stuff', he says. The routine for detecting these involved three steps. First, a map of the seafloor was built up with the swath mapper. Next, a CTD laden with sensors to detect the presence of a hydrothermal plume would be lowered. If one was found, the ship could then use a 'tow-yo'. 'This involves bringing the ship to a stop', says Richard, 'dropping the CTD apparatus towards the seafloor, finding the plume, then moving the ship forward at a speed of just one knot, and pulling the CTD up and down through the plume, a bit like a yo-yo, but reversed.' The *Surveyor* could then be swung around at right angles to examine the

Richard Arculus sits in the operations room. Source: MNF/Charles Tambiah.

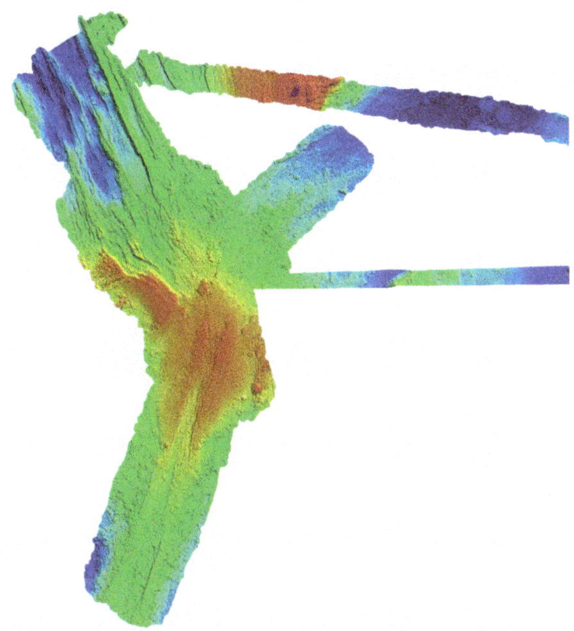

The triple junction between Fiji and New Caledonia, where three tectonic plates meet. Source: MNF.

plume's strengths and distributions, creating a two-dimensional portrait of its make-up. 'Then we'd put a dredge on the floor and pick up the rocks', says Richard. 'Then you just repeat the cycle: map, CTD, rock sample; map, CTD, rock sample … and so on.'

Richard's initial *Southern Surveyor* expedition in 2003, planned to begin from New Zealand, had a somewhat ignominious start. 'It had just been put back into service', he says, 'perhaps in a bit of a rush, and wasn't in the greatest nick.' The Auckland port authorities agreed and promptly issued the *Surveyor* with an unseaworthiness notice!

'It was like an unroadworthy for your car', says Richard, 'and we weren't allowed to leave the harbour.' The problem, it transpired, was the ship's drinking water, which had been coming out of the ship's taps in a rather unpleasant muddy shade of brown. Rust in the tanks had evidently been stirred up in the refit. 'It was perfectly safe to drink', says Richard, 'but it didn't look very nice.' After cleaning and refilling the tanks, the *Surveyor* was allowed to leave port. From this point, their luck turned.

North from the top of New Zealand lies the 10 km deep Tonga-Kermadec Trench. On its western side, ten active volcanoes poke their heads above sea level. 'Then there's nothing for 1000 km', says Richard. Scientists had often pondered the comparatively small scale of such an active system compared, say, with the Aleutian chain of nearly sixty active volcanic islands where, Richard tells me, a large and active volcano can be found every 100 km.

Soon after arriving at this supposedly empty sea to New Zealand's north, the *Surveyor* found that the few known volcanoes in this area were in fact not alone at all. Richard began to discover more of them, one after another, and in large numbers. 'It turns out, they're all underwater', says Richard. 'In the past ten years we've discovered 80 undersea volcanoes here. And the only way we could have found them was to go out in a ship like the *Southern Surveyor* and map the seafloor.'

There are only two occasions in which an undersea volcanic eruption has managed to be caught on camera – for years the holy grail of undersea geological exploration – and *Southern Surveyor* was directly responsible for one of them. It happened a couple of hundred kilometres south-west of Samoa at the north end of the Tonga Trench at a place called West Mata. 'We found it and swath mapped it for the first time', says Richard. Like many a great discovery, it was a complete accident.

'We were working out to the west doing seafloor mapping when one of the engineers wrenched his back working in the engine room, causing some serious and very painful damage', he says. The *Surveyor*'s captain, increasingly concerned about the ship's medical store running out of painkillers, decided to pull out of the expedition and deliver him into the hands of proper medical attention. 'The captain told me we'd be going to make a run for either Fiji, Tonga or Samoa', recalls Richard. 'In the end it turned out to be Samoa.'

Studying the chart of their route, Richard went to the master with a proposition. 'I asked him if we could take a particular route going there and another route coming back. I told him it would only add an hour to the voyage, but it would bring us across ground that nobody had ever seen before', he says.

Returning from the successful mercy run in a very deep 9 km of water, *Surveyor*'s swath mapper picked up a clearly defined conical-shaped ridge, rising up 5000 m from the seabed. 'We drove right over the top of it, mapped it and kept on going back to where we'd been working', says Richard.

The discovery and its location were passed to US scientists who, a year later, 'sniffed' the water column above West Mata, sent down cameras on a remote operated vehicle, discovered to their astonishment it was actually erupting and, for the first time, filmed it. The results, now on YouTube, are truly spectacular: massive, violent explosions of smoke, debris and flame; flows of orange lava pouring out in great underwater pillows which, at 1200° Centigrade, hit the near-freezing water and instantly fuse. Richard hands me a strange object, heavy and translucent, which resembles a kind of rough glass. 'You're holding one of the nodes of those pillows of lava', he says. 'The volcanic gases get trapped and stored inside it. Careful, it's quite sharp.'

He shows another piece of the remarkable footage of the exploding underwater volcano at West Mata. 'Isn't it fantastic?' he says with a rapturous air, and seeing it myself I have to agree. 'This is 1500 m below sea level. On the surface, you wouldn't

Superheated molten lava, ~1200°C, is about to explode into the water. The area in view is ~2–3 m across in an eruptive area approximately the length of a football field that runs along the summit of West Mata. Source: US National Oceanic and Atmospheric Administration.

notice a thing. Every time we go to sea we find new stuff like this. It's absolutely unbelievable.'

Richard would have needed every ounce of his palpable energy as Chief Scientist aboard *Surveyor* when, he estimates, he averaged about four hours' sleep a night for a month. I ask him, as I have others, how he managed to function. 'You get mentally fatigued', he tells me, 'but it's always so exciting you don't feel bad. There's always a new swath map, or new dredge of rocks or new hot springs, things that nobody has ever seen before.'

Time, I suggest, is a critical factor when conducting undersea exploration. I try to put myself in Richard's position, in the *Surveyor*'s operations room, watching the tantalising though incomplete images of a seamount or volcano coming up from the swath mapper and onto the screen, the ship essentially under his control as he must decide how to direct the bridge. Forwards? Backwards? A little to the side perhaps? It would be like trying to make sense of a vast tapestry in a darkened room with only the smallest of flashlights to illuminate the picture and understand the story. 'That's it exactly', he says. 'It's always a judgment call about what's likely to be down there and the best way to see it. The swath mapper just reveals it strip by strip and you're just guessing, "What's this? What's that? Is it a crater?" That's why I don't go to sleep because I don't want someone else making the call in case we miss something.'

For all his enthusiasm, Richard does not take the privilege, nor the responsibility in what he is doing, lightly. 'You've been given this tool by the Australian taxpayer and it's costing $50 000 a day to operate. I always feel a great deal of responsibility about running it as hard and as productively as you can', he says. 'It's like kids before

Christmas not wanting to go to sleep in case you miss something. The crew like it that way, too. They don't want to be sitting around doing nothing.'

Exciting or not, there are still times when the sheer exhaustion can start to gain the upper hand. Sometimes he'd send calculations up to the bridge and they'd call down, "Er, Richard, do you really want us to drag the bucket over the seafloor for *sixty nautical miles?"'*

I ask Richard about some stand-out memories, the wow factor or 'eureka' moments that stay with him. 'Oh, there are many', he says, not at all unsurprisingly. 'I'll just give you one.' About six years previously, he had been up in the north-west Laos basin. 'The map told us there was an edifice down there but none of the details', he says, showing me an impressive swath image revealing the unmistakable outline of a massive volcano. 'It was called "lobster"', he tells me. 'A massive caldera with these radiating ridges coming out of it.' The summit of this monster comes to just 600 m below the surface after rising a full 2 km from the seabed. 'It's just immense', he says quietly, poring over the coloured lines and patterns of the swath. 'And nobody had ever mapped it. Nobody knew it was there.'

After meeting Richard, I decided that one of the most impressive things about him was the way he still, even after many decades, manages to retain the sense of wonder at what he does. 'It's unreal', he says, with a shake of the head. 'It's a privilege to be involved in these voyages of exploration. And to have this facility, this scientific tool to do so.' The last ten years, he tells me, have seen a huge advance in our understanding of how the Earth works in the plate subduction zones, some of the most important places on the planet. 'The deepest trenches, highest mountains, biggest earthquakes, most explosive volcanoes are all associated with these places. We're discovering things all the time. Even in things like geology. You'd think there'd be nothing new to discover in geology but we have. We've done things like discover a new type of magma, for just one example.'

One aspect of his work – beyond lifting the veil on what lies beneath our oceans – speaks to a contemporary strain of public interest. 'You will hear people say', he tells me, 'when they see a volcano with a gas plume coming out of it, "Look at all that carbon dioxide, we don't have to worry about burning coal. Volcanoes put out more CO_2 than we do, it's natural." We know now that this is not the case, that their total carbon dioxide production is a hundredth to a thousandth of what we're doing ourselves.' He's full of surprises.

THE SCIENCE SUPPORT STAFF

Alicia Navidad

Hydrochemist

> *'The captain told me afterwards it was the roughest trip of his whole thirty-year career.'*

Alicia Navidad hasn't spent her entire sea-going career with *Southern Surveyor.* In fact, most of the time she's been intimately examining the composition of the oceans as a hydrochemist from within the big red walls of the Australian Antarctic Division's icebreaking resupply vessel *Aurora Australis*.

'The Antarctic voyages I did were usually concerned with climate change projects and we often went to similar places. But time on the *Surveyor* was amazingly varied: geology, currents, climate change, it was just so diverse. One time I even did some work with rock lobsters.' Although the migratory habits of rock lobsters isn't normally a subject Alicia would describe as being close to her heart, she nevertheless found something magical in collecting the tiny juvenile creatures – 'like bizarre tiny glass spiders', she says, echoing Kel Lewis' impressions of them. But what really floats her boat (as it were) is measuring the complex nutrients – nitrates, phosphates, silicates etc. – which are used to track the changes in the world's oceans, a task in which she prides herself on being very, very accurate.

After her first experience with *Southern Surveyor*, however, it's a wonder she ever agreed to set foot on the ship again. Freshly arrived from Adelaide in January 2006, Alicia's maiden voyage on the MNF was slated to be a relatively straightforward mooring trip from Hobart to Esperance under the aegis of the Chief Scientist Tom Trull, whom she was keen to impress. But that's a little hard to do when you're steaming straight into a massive low-pressure system in the Great Australian Bight where the sea is a mountain range and the waves are crashing right over *Surveyor*'s bridge.

'The seas were insane', says Alicia. 'I reckon I can count on my hand the number of times I've thrown up in my life, but on day three of this voyage, I remember just standing in front of my boss – my *new* boss – retching uncontrollably and thinking, "What have I signed up for? I can't do this!"'

A late larval-stage western rock lobster or puerulus which is being studied in an incubation tank in one of the laboratories onboard *Southern Surveyor*. The species is Western Australia's largest and most valuable fishery, and research into puerulus is critical to managing the fishery. Source: MNF/Alicia Navidad.

Simply moving about the ship became a serious problem. 'It got to the point where using the shower was out of the question', she says. The adage about safely manoeuvring inside a tossing vessel – *'one hand for you, one hand for the ship'* – was never learnt so well, nor so quickly. In all her many subsequent voyages Alicia never experienced anything like it, which is not surprising, because neither had anyone else. 'Everyone was saying that this was rougher than usual, but you don't know what kind of scale of "usual" they're talking about. The captain told me afterwards it was the roughest trip of his whole thirty-year career.' Despite appearing to have spent the entire trip hove-to in the storm, all the scientific tasks were accomplished, with every minuscule window of opportunity permitted by the furious weather being utilised to the full. But what a way to do it.

Alicia seems to enjoy throwing herself into her work, and never more so than when at its most intense. 'Many of the trips I was on were non-stop for usually more than twelve hours at a time', she says. 'You're just trying to keep up with the CTD casts, sometimes juggling three instruments at a time, always chasing your tail.' The more intense voyages she describes as her speciality: continuous CTD casts of twenty-four-bottle samples, one after another, up to five in a twenty-four hour period.

Sometimes she came close to pushing herself too far. 'The biggest trip I did on the *Surveyor* was three legs in row. That was madness', she says. A transit voyage, then two four-week deep ocean CTD voyages back to back kept Alicia at sea for sixty-one

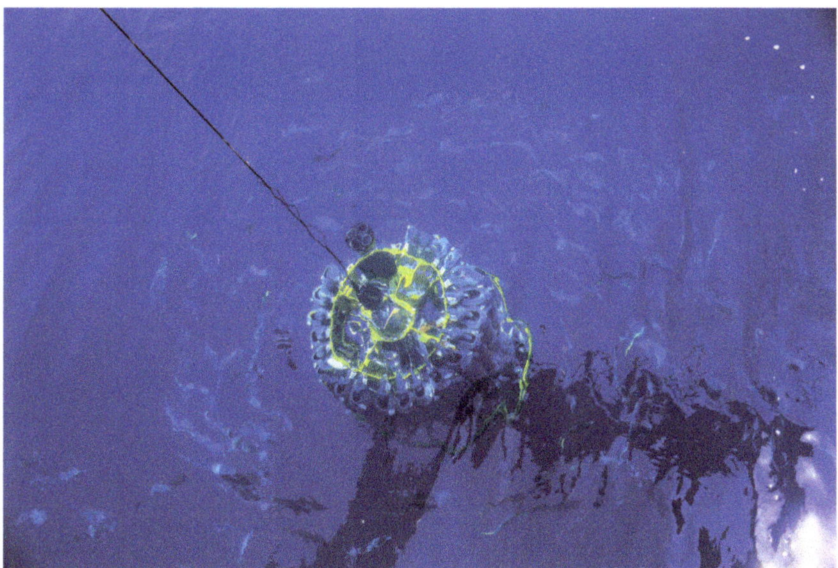

Tropical fish circle the CTD as it is lowered into the waters on a voyage led by Bernadette Sloyan into the Timor Passage and Ombai Strait, to study the Indonesian Throughflow. Source: MNF/Alicia Navidad.

days. 'And it was intense', she says, 'sixteen hours straight of nutrient analysis every day, and we were developing new protocols to achieve extremely high levels of accuracy with margins of error of less than 1%, so it wasn't something you could just turn up to, do your job and go home.' The voyage took Alicia and the five other scientists who, in her words were also 'mad enough' to sign up for this odyssey, from Brisbane to Wellington to Tonga to Lautoka. On arrival in Tonga, the fresh food started to run out. 'Something went wrong with the quarantine', she recalls with a slightly desperate laugh, 'so instead of all these lovely fresh fruit and vegetables we'd been told were coming up from New Zealand, all we could get in Tonga were bananas, a few eggs and some yams.' From what she remembers, most of the crew gave the yams a wide berth.

Even though this extended voyage knocked some of the stuffing out of her (she jokes about post-voyage trauma counselling, and being not quite so keen to put up her hand for the next trip), it helped establish her as a specialist in high degrees of data accuracy. Never did she feel unappreciated by the scientists she was working with, who, after all, depended utterly on the accuracy of the work she was producing. 'I'm a very meticulous person', she says. 'I take it on my shoulders to give the highest possible accuracy, because what we do to provide the best numbers possible to the scientists is incredibly important.'

But as is often the case, a scientist is sometimes only as good as the equipment they are working with, and sometimes that equipment can fail. With twenty-four hours' notice after a colleague was forced to pull out of a CTD voyage for medical reasons, Alicia stepped into the breach, finding herself right in the deep end aboard *Surveyor*, working for the first time essentially on her own, sampling as well as performing all three areas of CTD analysis. 'I was working very long hours, but it got to a point where I started to fall behind', she recalls. Waking up one morning, she distinctly remembers thinking, 'The only way I am going to catch up is if that CTD doesn't come back onboard today'. She was immediately wracked with guilt for even having thought something so heinous, so got up and prepared for another long day, starting with a decent breakfast.

Walking into the mess a little later, Alicia recalls one of the 'freakiest moments' of her life when in hushed tones a colleague asked, 'Have you heard?' Alicia looked blank. 'The CTD didn't come back up last night.' 'Oh … really?' she said, turning pale.

The loss of this highly valuable piece of equipment has been recounted by several of *Surveyor*'s crew, and explanations of what happened seem to vary like an old piece of folklore. Alicia believes the loss was caused by a sudden jolt of the extremely heavy rosette just below the surface as it was being retrieved, finding a weak spot in the slackened cable, snapping and free-falling to the bottom. I quip that it isn't strictly lost, that its location is more or less known so it's just a matter of going and getting it – which is what was attempted for the next two days. *Surveyor* dragged a steel claw back and forth across the seabed trying to hook and retrieve the quarter-of-a-million dollar CTD rosette. 'Instead, the boat got snagged', she says. Instead of hooking the rosette, the *Surveyor* found a rock ledge and was stuck for six hours as it gingerly attempted to disengage itself. Once free, the Master decided to count his blessings, cut his losses and abandon the hunt. 'I didn't wish for the boat to get stuck, I promise!' protests Alicia. I tell her I don't believe her for a second. After putting together the spare twelve-bottle rosette from scratch, it was back to work. But at least she'd caught up with the work, and didn't fall behind again.

Although ship life prevented Alicia from pursuing her passion for cooking and gardening, another hobby, photography, was easily indulged. Sea and landscapes, rare albatross and other images from the deck have all been caught by her lens in her spare time, but one shot in particular sparked for Alicia a degree of fame. When she describes the image, I immediately recalled seeing it on a banner in no less a place than the CSIRO foyer in Hobart. 'It's gone all over the world', she tells me. The image, of an Argo float being launched from *Surveyor*'s stern deck, has indeed travelled widely, recently appearing on a book cover and featuring in an exhibition in Paris. It is a fine shot, a study in action and concentration, caught at just the right angle at just the right moment as the float is hurled from the ship by the crew, their arms frozen in the moment of release, the expanse of the ocean providing the sweeping backdrop. Something about its composition brought to mind the famous Iwo Jima shot of US marines raising Old Glory atop Mount Suribachi in 1945, although Alicia blushes at the suggestion.

Alicia's image of an Argo float being deployed by crew members. Currently there are over 3000 Argo floats in the ocean, which are constantly collecting temperature and salinity data and relaying it back to scientists around the globe via satellite. Source: MNF/Alicia Navidad.

'I only managed to get it because I was hiding', she tells me in a slightly apologetic tone. On the day in question, Alicia knew an Argo launch was to take place, but not exactly when. While sweating through a session in the ship's gym, someone hurriedly announced, 'It's just about to happen!' Still in her gym clothes and being, in her words, 'a bit of a shy creature', Alicia raced to the back deck, camera in hand then, seeking a little privacy, climbed a ladder to the back of the dogbox (the winch control room) where she found herself close to, and slightly above, the aft deck from where the Argo was being launched. 'I just fired off, and one of the shots went viral', she says.

Alicia looks upon her time with *Southern Surveyor* as almost universally positive. Only once was she forced to abandon a voyage, when affected by a rather ghastly sounding mystery illness on a trip to the Barrier Reef. 'I broke out in these terrible throat blisters', she says. 'I couldn't eat, couldn't swallow, couldn't talk and I was in agony for days. But I'm a pretty determined, "don't give up" type of person.' Turning up to her shifts nonetheless, sometimes in her pyjamas, she had a note explaining to colleagues that she was, for the time being, mute. Eventually, a decision was taken on her behalf and she was off-loaded into the arms of her husband who had flown up to Cairns to retrieve her. 'I think it was an allergy, but the doctor never did find out', she says.

Southern Surveyor has allowed Alicia to see some extraordinary, otherwise inaccessible places, provided her with a career in an important and meaningful field and, via her images, has even helped spread her name abroad. Was there anything, I ask, she felt she didn't get to achieve or see in her time on the ship? She looks a little furtive for a moment, then confesses she'd always harboured a desire for a close encounter with the legendary Kraken, or giant squid. 'I didn't want it to sink us!' she says reassuringly, 'but perhaps for just a tentacle to appear out of the deep and touch the deck and disappear again.' I wonder how many of her colleagues share the fantasy.

Steve Thomas

Electronics technician

'The scientists would decide where they wanted to sample and we'd help make that happen.'

The first story Steve Thomas tells me about *Southern Surveyor* is about a bunch of blokes punching a hole through the bottom of her. 'They cut through the steel plate and then jack-hammered their way through the permanent ballast which was a mixture of concrete and steel balls', he tells me. 'It was a terrible, dirty job.' This is no tale of sabotage, rather one of how, in November 2003, one of *Surveyor*'s most innovative features was installed – the multi-beam seafloor mapping system, aka 'swath'. As an electronics technician Steve is, not surprisingly, the organised type and as we meet in the busy canteen at CSIRO's Marine Laboratories in Hobart he passes me some photographs he's brought along, plus a well-laid-out summary of his career and some highlights of his eighteen voyages with the ship. 'Putting this in was a game-changer for the vessel', he tells me. The image shows the *Surveyor,* oddly forlorn and vulnerable-looking, as all ships are in drydock, its pink hull exposed and its large bulbous capsule hanging below.

'That's the gondola', says Steve. 'It was welded onto the hull and is full of transducers for mapping the seafloor. They put a hole through the ballast and had to run all the cables through it. It took just over a month to complete in drydock in Fremantle, but it was a huge step forward.' Not only did the *Surveyor* have nothing like it, neither did any other ship operating in Australian waters at the time. 'It was a major change to the way the vessel was configured', he says. 'It's what allowed us to start mapping the continental shelf.' A year later, again in drydock, this time in Brisbane, Steve oversaw another valuable addition to *Southern Surveyor*'s exploratory arsenal, the sub-bottom profiler. This remarkable device, a transducer added to the gondola, enabled not only the seafloor to be mapped but also the layered composition of the seabed underneath, sometimes down to the bedrock.

Handling the pictures, I was curious as to why the famous swath seems to have been installed slightly lopsided, favouring the port side of the ship, rather than

The gondola fitted to the hull of *Southern Surveyor* was equipped with seafloor mapping and fish-finding sonar that operated down to around 3000 m. Source: MNF.

positioned dead centre. Steve gives a rather complex explanation as to how, in rough weather, air is drawn under the ship's hull and can be picked up by the transducers to give a false reading, and as air from the thrusters can do the same, the gondola is slightly out of the stream of interference. 'Oh, and there's another reason', he says as I prepare to digest another complicated slice of scientific logic. 'When we're in port, we usually park the ship on the right-hand side, so there's less chance of it hitting the wharf.' That one I have no trouble understanding.

He's also rather fond of the installation of *Surveyor*'s famous A-frame which picked up and retrieved so much of the heavy equipment off the back deck. 'We removed a gantry from when the ship was a fisheries vessel and installed this A-frame', he says. 'It was extraordinarily useful, enabling so many activities onboard the ship and was the envy of a lot of other research vessel operators because of its size and arc of movement.'

The 'bread and butter' of many of *Southern Surveyor*'s research voyages was the casting of the CTD, and a day in life onboard for Steve involved being in charge of every aspect of their deployment. 'If myself and my team were on shift, we were effectively in charge of that operation', he says. Making sure the CTD was ready to go over the side was a complex and physically demanding business. Steve and his team would start with pre-deployment checks on all the instruments and make sure the

The Autonomous Benthic Explorer (ABE) is loaded onto *Southern Surveyor* at the CSIRO wharf in Hobart. The ABE is a submarine that belongs to the Woods Hole Oceanographic Institute in the US. It was used to discover and map the location of fossilised corals south-east of Tasmania at depths of up to 4000 m. Source: MNF.

bottles were open and primed. 'The scientists would decide where they wanted to sample and we'd help make that happen', he says. Once 'on station' at the correct position, the ship would be told to stop, with the weather usually on the starboard bow. 'That was important', says Steve. 'If the weather picked up, it could push the vessel over the wire as it was travelling down and damage it.' Much like a fish dragging your line under the boat, I suggest.

Steve would communicate over the intercom with the winch driver, knowing they were ready to pick up the CTD, put it over the side and start lowering it. 'We'd tell them how deep it had to go, usually to within 10 or 20 m of the seafloor', he says, although care had to be taken, if the ship was rolling, not to slam the heavy CTD onto the bottom. The CTD would descend at 1 m per second, the limit set by the electronics of the instruments to record temperature and conductivity etc. on the way down. 'If you went too fast, you'd end up with gaps in the data', he says.

On the way back up, the scientist in the operations room would decide the precise moment for the rosette's bottles to snap shut, capturing a sample of the ocean at a given depth. 'We'd be in constant contact with the bridge and the winch driver to let everyone know how things were going', says Steve. Once the heavy instrument was back onboard and its samples taken, Steve would advise when it was time to move

The CTD is deployed during a voyage from Cairns to Gladstone in 2008, where data was collected on the uptake and storage of carbon dioxide throughout the Great Barrier Reef. Source: MNF/Bob Beattie.

on to the next station for another cast. Depending on the depth of water, a CTD deployment could be executed every hour.

The variety of instruments which could be attached to the CTD, all under Steve's aegis, was extensive, depending on the particular secrets scientists wished to extract from the deep. 'Sometimes they'd like to put on fluorometers to measure the chlorophyll which can gauge biological activity in the water, or transmissometers which allow particles to pass through a light beam to measure sediments.' There's also something called an ADCP (Acoustic Doppler Current Profiler) which uses the Doppler effect to measure the direction and velocity of currents. All these instruments, ranging in size from the size of a coffee cup to 0.5 m long, would have to be enclosed in a strengthened stainless steel or titanium casing to withstand the gigantic pressures, attached to the CTD and sent to work. I ask Steve if we make these instruments ourselves. 'We used to', he says, 'now most of them are bought off the shelf from overseas.'

Things were rarely lost – one CTD in ten years, according to Steve – but whenever something did it warranted the 'Yo Award', 'As in a yo-yo that goes down, but fails to come back up again.'

Having given me a picture of a typical working day when everything went right, Steve then proceeded to tell me about a couple of trips that were, for various reasons, hellish.

'Poor Dr Neville Exon', he begins, with a shake of his head. 'That trip was a whole series of failures.' The Kenn Plateau lies north-east of Brisbane and in 2004, with a seismic system, Neville intended to map some of its deep and spectacular terrain. A seismic system can reveal the composition of the underseafloor to astonishing depths – up to 1 km, Steve tells me. Using 'air guns', high-pressure air bursts of 2000 psi create low-frequency acoustic signals, the returns of which are measured by a trailing series of hydrophones. The information is read and interpreted in the *Surveyor*'s operations room. Ideally, mud and sediments laid down over millions of years between seamounts and in canyons – as well as the geological history of their underlying bedrock – can be revealed by this remarkable device. That is, when it all works.

'This area north of Brisbane hadn't been mapped before so it was quite an important expedition', says Steve. 'Unfortunately, the compressor they were using wasn't reliable and failed fairly early.' Having used the spare parts they carried, new parts had to be ordered, flown up and delivered by charter boat, at sea, several hundred nautical miles off the coast of Bundaberg. 'Then we needed more spare parts', says Steve. 'Five days later they were delivered, courtesy of an air drop by a charter aircraft!' The parts had to be wrapped in plastic foam, glued tight into a PVC pipe and packed into an emergency liferaft box. Arriving above the *Surveyor*, the plane did a couple of circuits then shoved the whole thing out the back. It floated down on a parachute and was retrieved by one of the ship's launches.

'Things then got back on track – for a while', says Steve. Soon after, the *Surveyor*'s hydraulic return line leaked then failed completely, requiring all the ship's winches to be shut down. Having been planning this trip for a good couple of years, Dr Exon was, at this stage, says Steve, 'starting to run out of options'.

Where can you get a winch fixed in the middle of the Coral Sea? A quick look at the chart led to the *Surveyor* being turned around and steered to New Caledonia, just in time for a long weekend during which every tradesperson in Noumea high-tailed it out of town. The repairs proved to be a far bigger job than anticipated and the impromptu stay in Noumea ballooned to several days while the recalcitrant hydraulic pipe was located and repaired. I suggest there are probably worse places than Noumea to be stuck but Steve doesn't think Neville would have seen it that way.

The tale of this trip pales, however, compared to what transpired on a 2005 voyage to the Gulf of Carpentaria with geologist Dr Peter Harris, when a blockage occurred in the vent pipe of the ship's sewerage system. 'The hydrogen sulfide or "rotten egg" gas built up quickly', says Steve, 'and vented itself back out through every toilet in the ship. Actually', he adds on reflection, 'I don't think I want to go into too much detail about that one.' I had eaten recently, but press him for a few more details. 'Well, we had to evacuate the ship and come out onto the deck to escape the smell. The gas from the sewage builds up incredibly quickly and without anywhere to go, well, it had to go somewhere.' Pipes on ships are often virtually impossible to locate, buried behind bulkheads and decks during construction. 'It took the engineers about a week to find where the problem was', says Steve, 'and it was a couple of weeks before it was fully fixed.' Corrosion in the old pipes was the cause, and although it didn't impact on the voyage's work it was not a pleasant trip. 'I can laugh about it now, but it took us a while!' he says.

On another voyage Steve would rather forget, absolutely nothing failed and the ship behaved perfectly. The same, however, cannot be said for the Chief Scientist. 'She had two PhDs', he says. 'I thought it was a typo when I first looked at the voyage plan.' But no, this academic's title really was 'Doctor Doctor' and she insisted on it being used by everyone onboard. I tell him it sounds more like something from a TV sitcom, but Steve assures me it was true.

It stemmed, Steve tells me, from a personality clash between the Doctor Doctor and her deputy, who had brought along a group of students and was seeking a small amount of time to show them some aspects of marine science as experienced at sea. 'The Chief Scientist was determined not to let that happen', says Steve. This was her voyage, and hers alone. 'The students became despondent, the deputy became furious, and the bridge began to question why they were getting confused and muddied instructions.'

At the end of the voyage, the Chief Scientist relented and allowed the students to take part in a CTD deployment – and the winch broke down. Steve could not believe it. 'It just capped off a very stressful voyage for me.'

The moments Steve prefers to remember are the many voyages which achieved their aims, when the ship behaved herself and the scientists were happy. 'What made it special for me were the times you came off the voyage and the chief scientist had achieved everything and more than he or she set out to, and were singing the praises of the ship and the crew. That's when you knew you've made a positive contribution to a science program. It's not easy out there most of the time', he says. 'But in the end you know that you're doing something that's never been done before.' It's one of those statements which, as we our return our coffee cups to the trolley in the CSIRO canteen, brings home the amazing reality of science. I thank Steve for his time, and as he turns to go he remembers something. 'Oh', he says, 'and dolphins riding the bow wave. Seeing that was pretty good too.'

Matt Sherlock

Electronics engineer

> *'Scientists are always asking for things that are impossible, but after a while at sea you know what's going to work and not work.'*

Electronics engineer Matt Sherlock goes way back with *Southern Surveyor*, to when the vessel was first bought by the CSIRO Fisheries Division from the Norwegian company which, for a reason no one can figure out, named her *Southern Surveyor*. She looked quite different in those days, says Matt, even down to her livery. 'I remember when she arrived in Australia in 1988. She was painted orange with a white superstructure', he tells me. She also sported an extra 'fly' bridge and a large 4 × 5 m 'moonpool', open to the water from inside the hull in order to protect divers from the waves when the ship was a dive support vessel. 'Actually, she was a little tired when she arrived', he remembers, but her year-long refit in the Launceston shipyards, in which Matt had a part to play, changed all that. New winches and science laboratories, a streamlined bridge and a nice coat of her famous dark blue. 'We didn't get rid of the moonpool entirely', adds Matt. 'We just made it a bit smaller.'

Matt has lost track of the number of voyages he's completed on *Southern Surveyor*. 'Gee, I dunno ...', he muses. 'Twenty-five?' However many it was, you can bet not one of them was a doddle. Looking after the scientific equipment – some of which you've made yourself – for a three-week marine science voyage involves a good deal of time spent on the ship's open back deck preparing gear, making sure the connections are right and the systems are running. 'It's an excellent role to have', says Matt, 'and very closely involved in the achievements of the voyage.' But that back deck can be a hazardous place, and one afternoon Matt very nearly left a piece of himself there as testament to its dangers.

'Scientists are always asking for things that are impossible, but after a while at sea you know what's going to work and not work.' For example, he says, deploying a piece of equipment weighing nearly a ton can be relatively straightforward in calm weather but in heavy seas it's a different story, particularly when it comes time to retrieve it. 'When the ship's rolling', says Matt, 'and if it starts swinging on the end of a cable off a gantry or an A-frame, it can do a lot of damage – to the ship and of course the gear itself. That's where practical experience comes in.'

It's not just the weather that could make the *Surveyor*'s rear deck a dangerous place to work. Sometimes it simply gets too busy. 'There can be so many activities happening there all at once it gets very confusing, particularly at two in the morning', says Matt. 'You might have a fibre-optic cable on the deck and someone might be moving a two-ton dredge and accidentally drag it over the cable and damage it, then you have to re-terminate it, which is a six-hour job. Those sort of frustrations happen. It's just the risk of doing things at sea. Something's eventually going to go wrong.'

Nothing, however, went wrong on *Surveyor*'s very first trip out of Launceston Marine's yard after its major refit as a fisheries research vessel, and Matt remembers it well. 'We came out of the Tamar under this absolutely clear blue sky, sun shimmering on the water, and followed a course into Bass Strait', he says. On the way to Hobart, it was decided to test the acoustic equipment on a seamount off St Helens, but they wouldn't be alone.

This was the height of the orange roughy fishing frenzy, and Matt recalls having to dodge virtually an entire fishing fleet. 'It was a famous place for the fish, and there must have been thirty-five trawlers out there all waiting their turn, for their "shot on the hill", as they called their trawl of underwater hill.' The practice was that boats which were waiting would leave their lights on and their motors running, and the crew took the opportunity to grab some sleep. 'We had to follow this careful zig-zag pattern in and out of them in order to test our acoustic gear to read the biomass. It was an incredible thing, cruising in and out of these fishing boats and looking down onto them from above', he recalls.

On a voyage not long after this, a serious problem with the ship's steering occurred. Matt was on that one too. 'We were out at sea and all of a sudden, we started going in circles', he recalls. 'I couldn't help thinking, what if this had happened back when we were among those fishing boats or close to shore?' The hydraulics in the steering system – flagged as a problem when the vessel arrived from Norway – were replaced, and not before time.

Matt also accompanied the *Surveyor* on an infamous voyage from Hobart to Western Australia and, unlike many I have spoken to who sailed in her, freely confesses to suffering the scourge of seasickness. 'It was the worst trip I've ever been on', he says plainly. 'Don McKenzie and I were onboard, and it was September. We went south-about (around the bottom of Tasmania) and straight across to WA and I basically threw up for ten days.' Ploughing into 10 and 20 m waves with the spray coming up to the front windows of the wheelhouse, Matt began to think about the dangers this presented to the electrics. 'We opened up the consoles where the electrical switch gear was housed in the forward control units and there was two inches of salt water sloshing around in the bottom of them.' This could easily short and stop the steering or even the engine.

One day during this hellish trip, all they could do was turn hove-to into the gale and wait it out. 'With the electrics in mind, I remember asking the skipper, Jim London, what would happens should the engines fail. He turned to me and said quite dryly, "Well, we'll turn sideways, roll over and all die".' Sobering words, but at the time Matt was grateful as hearing them shocked him out of his seasickness.

Matt eventually found his sea legs and learned the hard way that the *Surveyor* was built for weather. 'A cargo ship off WA was severely damaged in that storm, and at one stage we had an Orion air force plane circling overhead asking if we were okay, but she's an old North Sea trawler and incredibly strongly built.' The only time I saw *Surveyor*, when given a tour by Mike Jackson, he pointed out the strength of her bow. Her close steel ribbing certainly looked built to withstand anything.

'So, tell us about the accident', I asked Matt. I had already heard the story from other crew members and scientists, particularly Boysey, who played a part in the drama, but I wanted to hear it from the man himself. 'Well, it was a funny thing, really', he muses, involuntarily flexing his right hand as he begins the story, without giving a sense that he's being made to drag it out for the hundredth time. 'We were almost down at Macquarie Island to do underwater camera work and take samples with the sled', he says, passing over a photocopied picture of a 'mini-sled' he worked on, complete with robust-looking rails that are dragged across the deep ocean floor. I make a suggestion that perhaps wheels would work better, which on reflection seems rather foolish, for any number of the reasons Matt politely gives me. 'But that's not the gear that did the damage to me', he says.

An acoustic towed body is a heavy cylinder-shaped object to which acoustic transducers (sonars) are attached in order to gauge the size of vast schools of fish. 'If you're working in deep water, echo-sounders on the hull spread the acoustic signal out too much and you can't resolve the fish stocks properly', says Matt. 'We took the transducers and put them on a 2.5 m torpedo and towed it about 1 km deep above the fish. If it's calibrated properly, you can read them down to the ton.'

After a successful deployment of this acoustic torpedo, Matt was engaged on *Surveyor*'s back deck, assisting the recovery as the torpedo was carefully winched onboard. 'On the deck there was a cradle that it sat in', he says. 'We had ropes attached to it as it was being lowered, but it didn't sit in it cleanly.' The ship was rolling and the torpedo's position was noticed by the deck crew. The winch driver was signalled to take up some of the weight, which would allow the crew to adjust it to sit properly. At this moment – and not wearing gloves – Matt happened to have his left hand on a steel handle attached to the cradle, with his middle finger slightly protruding over the side.

As the ship moved, the torpedo, nudged ever so slightly, found its way to a snug fit inside the steel cradle, coming to rest with a heavy metallic thump. 'Unfortunately', says Matt, about to say something I know will make me wince, 'my finger got jammed on the inside of it, as the torpedo came down into the cradle.' He says he certainly felt something, but his initial thought was that he'd been lucky and had gotten away with it. 'But then I looked at my finger and the whole top of it was gone and the bone was sticking out.'

His first reaction to the missing digit? Panic? An agonised shout? Some indication of stress or shock? Not a bit of it. Just the vague notion, he says, that 'this is going to be a bit of a problem'. So casual in fact was Matt's reaction that no one initially knew anything was wrong with him. He simply made his way to the ship's hospital and showed what was left of his hand to the Chief Mate and Medical Officer, Boysey. Somewhat ignominiously, Matt's missing finger had been trampled on then kicked to the other side of the deck, most likely only just avoiding going over the side and ending up as a snack for a Southern Ocean mackerel.

Surely, I suggest, he was at this point screaming in agony? 'Actually it wasn't too bad', he says, with a little surprise. He admits he was probably in shock, but suggests that the ripping rather than cutting nature of the injury lessened the pain. It wasn't a question I felt any desire to dwell on.

'Boysey tended to smoke and his hands were shaking a bit as he put in the stitches', he says, so he politely suggested that he would prefer to perform the operation himself. Boysey complied with his patient's wishes, prepared the sutures then Matt carefully reattached his own finger with three deftly applied stitches – all without the aid of anaesthetic.

'I think the secret for me was getting it straight back on', he says, although Boysey has claimed that his careful placement of the finger the right way around assisted in its eventual restoration! Whoever was responsible, there isn't much now to show that it was once a severed digit – it appears almost completely normal, with just a pale thin scar. Matt tells me that most of the movement and feeling has returned as well. 'It was just like sewing on a button, really', he says. The pain did indeed arrive an hour or two later, but painkillers took the edge off until they came ashore two days later at Macquarie Island to present before the somewhat surprised doctor. 'By this time the whole top of the finger had gone black, so I was rather concerned.' The doctor, after carefully removing the bandage, gave it a thorough inspection. 'Actually, that looks pretty good', he said, and so it turned out to be.

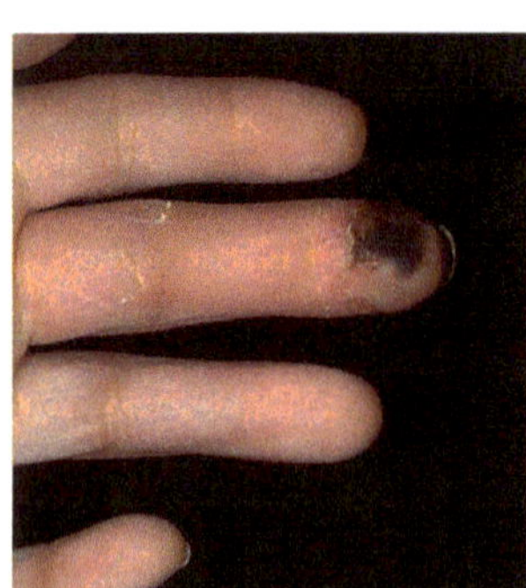

Matt Sherlock's finger a few weeks after returning to shore from the voyage to the Southern Ocean. Source: Matt Sherlock.

Matt stayed with the voyage for another three weeks, even continuing to work. Looking at his finger again, it's hard to believe the injury was as serious as it was. 'I was amazed myself', he confesses, looking at it as if it were a rare object behind glass.

If Matt's misfortune was a major one, it pales beside the near-miss of another member of *Southern Surveyor*'s science team, fisheries biologist Cathy Bulman. Matt was a key witness to this drama which unfolded in the mid 1990s, when *Southern Surveyor* belonged to the CSIRO's Division of Fisheries, on a voyage along the south coast of Tasmania. It was a normal day and the sea was fairly rough, but not so severe as to halt the scheduled CTD deployment which was done, in those days, from the open CTD laboratory on the side of the vessel.

'Back then we would drop them down the side of ship into deep water', says Matt. 'The ship would turn sideways to the sea when we deployed them.' It also meant that the ship had a tendency to roll, as it did in rough weather. 'I remember we were preparing to deploy the instrument and I'd just walked through the CTD lab and was going up a ladder to the deck above when the ship took a massive wave', he says. Gripping that ladder for all he was worth, Matt looked back down in time to see a great wall of green water rise up and engulf the small space of the laboratory through its open 2 m wide doorway.

'This huge wave just crashed into the lab, putting it under 1.5 m of water', he says. 'I watched all these toolboxes and chairs and things washing out and onto the "hero platform" below.' In the laboratory too was Cathy Bulman, who was instantly swept off her feet by the sudden deluge. As she headed out the open hatch and to heaven knows what fate, one of the able seamen, Mick, managed to get his hand on her jacket and haul her back. 'It was such an amazing thing to see that wall of water pouring back out of the lab with all that equipment', says Matt. 'We always used to

Matt Sherlock (front) and Rudy Kloser set up in the fish laboratory waiting for the live data feed from the towed camera to begin. Source: MNF.

In 2008 a voyage led by Alan Williams captured the first images of deep sea corals that covered Australia's largest cluster of seamounts, located south of Tasmania. The Tasmanian Seamounts are home to unique species of corals and sponges, some of which are hundreds and possibly thousands of years old. It is a significant habitat for orange roughy. Source: MNF.

joke to Cathy that Mick saved her life and so she had to marry him. And a couple of years later she did!' Apparently the water was not the only way in which Cathy was swept off her feet.

Matt has, in his time with *Southern Surveyor*, seen the odd failure – a rare cable snapping and a camera mount not coming back, and was as downhearted as anyone by the infamous loss of the SeaSoar, but overall the successes have spectacularly outweighed the misfortunes, including the near-loss of his finger.

'All through my career', he says, 'I've been surprised at how successful we've been. I think that comes from the engineering team who build this stuff with so much insight.'

He cites his proudest achievement as the innovations in deploying undersea cameras. In the beginning, high-definition cameras would be attached to packages which were towed blind at a certain altitude above the seafloor. 'We didn't know what we were getting back then', says Matt. 'We'd get these cameras, open them up and look at the tape. I remember surveying that seamount off St Helens and then looking at that tape and seeing these orange roughy 1 km deep, beautifully illuminated. It was such a buzz to see it.'

With the ability to view live, Matt could move a small joystick in the control room and witness, in colour, 'entire zones of the ocean that had never been seen before'.

Matt feels privileged to have worked with scientists who have been willing to continually push the technology to achieve the best results. 'I feel we do really well in a lot of areas, particularly in Australia with the limited budgets we have', he says. But it's a story that involves no technology whatsoever that he leaves me with. Once again, the stage is the *Surveyor*'s always dramatic back deck.

'The one thing that still sticks in my mind after ten or so years is one evening when I was out on the back deck by myself. We were cruising along with the sun setting in front of us, and big black storm clouds behind us, and an albatross, its white wings illuminated in the bright sunlight with the dark clouds behind it flying alongside at my eye level in the air wake of the vessel. I sat there just looking at it – the bird and I, just a couple of metres apart – and thought, "Wow, this is fantastic".'

Ian Hawkes

IT manager

'I was familiar with sea-going software, but I hadn't actually been to sea.'

At first meeting, Ian Hawkes seems far too gentle a person for the rigours of a sea voyage, being thrown about inside *Southern Surveyor* on the wild grey-green seas of the Southern Ocean, and after his first trip in 2006 – from which I strongly suspect he is yet to recover – he could only agree. But don't let the gentle exterior fool you. Previous to working with the MNF, Ian was involved in defence contracts, and even had a spell in Canada working out air traffic control systems. 'Not much pressure there', I remark, to which he laughs.

Ian is an IT specialist, charged with keeping the ship's complicated onboard software up and running and problem-free. Not an easy task in an air-conditioned city office, much harder in an ageing research vessel headed into the middle of nowhere. 'The problems are often the same as they are onshore', says Ian, 'but the working conditions are a lot harder, especially if it's rough and you have to hold onto your desk to stop you falling off your chair.' He was initially brought in to drag *Southern Surveyor*'s outdated onboard IT systems kicking and screaming into the 21st century. I quip that perhaps they were, after all, quite happy running DOS, at which he pauses and says, quite seriously, 'Actually I think there were a couple of DOS machines when I arrived. Also a couple of Sun machines. You don't see them around much these days either.'

Soft-spoken computer geek that he might be, Ian was nonetheless sent – literally – to the deep end upon his arrival. 'I went out to sea quite early', he says, and there is a pause before he takes a gulp from a glass of water, 'on one of the Tom Trull trips'. I can almost feel the shudder as he utters the words.

'I'm not your classic voyager type person', he confesses. 'And Tom's trips usually only went in one direction – south.' Having spoken to Tom, who freely confessed that his Southern Ocean mooring voyages were the ones 'nobody liked going on', it came as no surprise to hear Ian mention a saying that did the rounds of the MNF: 'If you can get a person on one of Tom Trull's trips for their first voyage, it's great fun for everybody onboard … except that person.' This would particularly apply if, like Ian, your only sea-going experience had been the Tasmanian ferry crossing of Bass Strait. 'It was a bit of an eye-opener', he says, with a nervous chuckle. 'I was familiar with sea-going software, but I hadn't actually been to sea.'

But off he went, bidding farewell to his wife and young children for a week on the waters 'down south' on a voyage to service Tom's infamous deep sea moorings. Barely one day out of Hobart, the suffering commenced. 'It was extremely rough – 15 m seas', says Ian. 'I recall going up to the bridge and looking out of the window. It was … horrific. Waves smashing against the windows … I honestly had no idea what was normal, or whether a ship could actually take these sorts of conditions.' He speaks of it like a soldier having come through a battle.

Although he says he felt like he was on a roller-coaster, Ian's modesty hides a resilience that even the most senior members of the crew found hard to match. 'I actually went okay. A little queasy, but I didn't get ill.' Not like the *Surveyor*'s Master at the time. 'A cool guy', recalls Ian, 'with a very adventurous spirit, but he always got seasick. You just don't know who's going to get affected or in what way.'

Days into the voyage, the conditions deteriorated and stories of a massive low waiting for them, somewhere over the invisible horizon, began to circulate ominously around the ship. 'We were having to heave-to and point the bow into the waves in

The operations room was the heart of the science operations, where data was received and stored until the ship returned to port, where it was downloaded. All data collected onboard the MNF research vessel is publicly available and free to use. Source: MNF.

order to survive', Ian says. Adding to their woes, the swell was forcing the *Surveyor*'s prop out of the water, overspinning it and making the engine automatically shut down, cutting off the power. Some of the bridge furnishings were beginning to fall apart. 'There was actually an announcement over the PA at some point that the bridge was starting to disintegrate', he says. 'I'm still not sure if the guy was serious.' Eventually, thankfully, the trip was aborted and *Surveyor* made a run for the safety of Adelaide.

Beanbag chairs were liberally spread around the operations room and when things were rough these were put to good use by the fading scientists and crew. 'The ops room was in the middle of the ship and quite low down and about the most stable place to be', he says. When things got nasty outside, 'bodies would start appearing, lying in beanbags and looking depressed'.

When on shore and communicating with his team onboard to solve some IT problem, Ian could usually tell when things were becoming difficult by the deteriorating quality of their communications. 'It can be physically and mentally draining', he says. 'You could see the state they were in because suddenly there'd be a lot more typos and they'd start saying things that didn't really make a lot of sense. Almost like they've been oxygen-deprived.'

Aside from the ship's engine, it's hard to imagine a component more vital to the running of the ship than its computer system. 'The whole purpose of the ship going out there is to collect data', says Ian, 'so it's extremely mission-critical that we run and monitor the logging systems for the instruments to make sure they're getting good data. If anything's going wrong, we've *got* to fix it.'

As well as the science and data-gathering, there are the crew's and scientists' personal computers and the ship-to-shore via satellite communications, which Ian had to keep running despite the challenges of salt, water and motion. Sometimes jobs required him to fly to a port where the ship happened to find itself; he would disappear into the bowels from dawn till dusk, or until the *Surveyor*'s schedule determined it had to leave. 'Port period visits were always pretty stressful', he says. Difficult jobs and immovable deadlines often collided. 'We'd sometimes finish something vital just a couple of hours before she was to sail.'

I ask him if the system ever crashed. 'Oh yeah!' he replies with a sort of gusto. 'That's why we ran two systems in parallel. We liked to pride ourselves on being able to fix something, or work around it. It was actually very unusual that you'd have a problem which actually prevented the ship achieving the objectives of the trip.' Having said that, he stresses that not everything worked all the time. 'It's a very competitive process to get time on the ship', he says, 'so the scientists are under a lot of stress to fit in as much as they can in a short space of time. Things don't go to plan very often. It's a ship, you've got weather and anything can happen.'

I ask him if he felt safe onboard that first trip. 'Not really, no', he tells me with candour. 'But that was my inexperience.' His only preparation for how bad conditions were going to be was via the ribbing of his more sea-seasoned mates. 'The old hands always delighted in telling you how tough the conditions were and how rough it was going to be', he tells me. But in this case, I dare to suggest, weren't they pretty much right? 'Well … actually yes they were', Ian admits, chuckling again. 'But you have faith in the crew, and you have faith in the ship. I met people on that trip who were also on their first voyage', he says. 'Some of them I'm still close friends with. Being on a ship can be quite a bonding experience.'

Jeff Cordell

Electronics support technician

> *'You build it up in the workshop, and it all looks fine, then when you get to sea you think, "Oh, that's not quite right".'*

For Jeff, one or two voyages that fit very neatly into the latter category occurred early in his career as one of *Southern Surveyor*'s science support technicians. The ship herself was pretty new too, at least in its Division of Fisheries incarnation. 'It wasn't long after her refit to become a fisheries research vessel and everyone was still learning, so it was all pretty rushed', he says. One of those trips he'd probably rather forget took place off the east coast of Australia, near Eden. They weren't long at sea when a water pipe burst somewhere near science cabin 1, drenching the most forward cabin on the *Surveyor*'s starboard side, then quickly spreading all the way aft and making a deposit in just about every room in between.

'I remember going down into the engine room where I was pretty shocked to see water pouring out of fittings right next to the main power switchboard!' But apart from reminding himself to take a torch to bed that night as he thought he might be needing it, there wasn't a great deal he could do. 'There was water coming through all these welded deckheads. We couldn't reach them, and we couldn't just stop the boat and mop it up. You've got to just wait for it to come out.'

At this point I remind him that one thing that tends to be scarce in the middle of the ocean is fresh water. Might not an incident such as this have consequences for the ship's precious reserves?

'Well, yes it did', he says. 'We had two reverse-osmosis water makers, but they didn't have the capability to make as much as the crew and scientists were using.' Besides, these ingenious devices relied on the main engines running at high speed, and with the ship working in a slower, more ambling mode during much of the scientific work, they proved not to be a great deal of use.

If that trip was a rough one, it was nevertheless a luxury cruise compared to the first voyage the *Surveyor* undertook after her 2003 refit. 'We left Hobart for Auckland for a twenty-one-day sail with scientists from the University of Tasmania exploring the geological features of the seabed', Jeff recalls, before pausing to do a slow, ominous shake of the head. 'The ship was still recovering from the major refit and things, well, let's just say, weren't quite as finished off as we would have liked.' This proves to be an understatement of near-Biblical proportions. On day one a pipe burst, and this time not the relatively benign fresh water failure of Jeff's earlier voyage. It was a ruptured effluent pipe which very quickly dumped a ton and a half of raw sewage into a forward bilge. 'We had to pump it out but it was full of scientific equipment which we then had to clean and sterilise. It was … er … not pleasant', he says. I believe him. But that was only the beginning.

Soon after, *Surveyor*'s steering mechanism began to play up, causing enormous stress for scientists and crew alike, and it had to be monitored constantly. Then, not one but both the reverse-osmosis water makers failed, forcing everyone onboard to rely solely on the ship's limited fresh water reserves. 'In the last week we were getting into the dregs of our holding tanks and the water started to come out brown', Jeff remembers. To cap things off, there was the weather: unrelentingly horrible from start to finish. 'It never blew below thirty knots the entire voyage', he says. 'We were constantly rolling. It was difficult to work, hard to sleep and impossible to relax. I remember walking down that gangplank in Auckland thinking, "I'm not even going to look back." And I didn't.'

The *Surveyor* was already several decades old when the MNF took her over. Jeff reminds me that an old ship is like a lottery: 'You never really know what you're going to get or when something is going to blow up.' On one occasion during the early days, things did blow up rather seriously. 'We were in Torres Strait and started to hear this funny noise whenever the ship turned, and couldn't work out what it was. Not a big noise, but little bangings, little like plastic floats hitting a tin can', he adds with the precision of a technician. 'I actually thought it was coming from somewhere midships.' The mystery sound was eventually traced to deep in the stern and a diver was sent to investigate. His ominous observation – 'scrape marks on the nozzle' – caused the engines to be immediately shut down and the Master to leap onto the radio.

It was found that the bearing holding the Kort nozzle – the large cylindrical device surrounding the propeller which turned to direct the prop's jet of water to steer the vessel – was loose. 'Wobbling', in fact. The loose nozzle, as it turned, was hitting the propeller, chewing out large metal chunks. The moment they heard this news, Jeff and the entire crew knew that the voyage was over. 'If that bearing had failed, it could have taken out the prop as well as the shaft and caused catastrophic damage to the ship', he says. The only thing to do was for the Master to hit the airwaves and call for a tow.

The next ten days, as they were hauled at an agonisingly slow two knots from the Torres Strait into Cairns, passed like an eternity. 'There we were', says Jeff, 'stuck onboard a slow-moving boat with no work and nothing to do for ten days. The food started running out just at the point when you suddenly have more time to eat it.' I can picture the science team, glum, having waited perhaps years for this trip, their work and expectations suddenly in ruins. Were they able to get that time onboard the *Surveyor* back? Were they entitled to go to the front of the next queue to use the vessel? 'Sometimes it was possible', says Jeff, 'sometimes it wasn't.'

Having sailed on both her first and her penultimate trip in late 2013, Jeff's sea time almost perfectly bookends *Southern Surveyor*'s life with the MNF. Most of his voyages were trouble-free, some memorable, others even spectacular. I ask him for one that stands out in ways other than injuries and broken sewage pipes. He hesitates barely a second before saying, 'Macquarie Island', and a large, quite mystical smile spreads across his face. Although even that trip was not without drama.

A keen recreational sailor, in January 1999 Jeff had just recovered from one of the most terrible maritime ordeals imaginable – captaining a yacht in the infamous Sydney to Hobart yacht race of 1998, when a deadly storm engulfed the 115-vessel fleet off the south-east Australian coast, forcing all but forty-four to retire early and taking the lives of six sailors. 'The worst experience of my life by a mile', he says. 'And it wasn't like a near miss that goes away, this was eight to ten hours seriously thinking we could well perish. I've never been so frightened in my life.'

So, for Jeff, the prospect of a trip deep into the Southern Ocean on *Southern Surveyor* just a week or two later didn't exactly fill him with enthusiasm. 'It was to be a thirty-day voyage to study the Patagonian toothfish, right on the edge of the boat's endurance', he says, and it would take him close to the wondrously remote Macquarie Island. Although the crew was not scheduled to go ashore, the view of this great wind-swept southern sentinel promised to be unforgettable.

All was going to plan until the moment his colleague, Matt Sherlock, came in off the rear deck with his missing finger. 'No, he didn't look real good', says Jeff, chuckling now, though I doubt if anyone was doing so at the time.

'We weren't planning on going ashore, but we were granted permission because of Matt', he says. 'Macquarie Island was the most pristine environment I'd ever been to. Never seen anything like it: being able to get up close to elephant seals lying there, soaking up the sun; amazing numbers of penguins which believe, me, you could smell as well as see – just like on a nature film but the real deal.' Matt's misfortune had enabled the *Surveyor*'s crew to have an impromptu visit to a place they would never otherwise have been able to access. 'We talked to all the scientists wintering there, who were happy to see another face. It was wonderful. I'd gone to being apathetic at the start of the trip, to having a life experience I'll never forget.'

Even the weather – grey, but not particularly heavy – was as kind as it can be in that remote and wild part of the world. On some trips, Jeff remembers it being far less kind. 'In seventy knot winds one time I was thrown 3 m sideways out of a bunk. I assume it was a big wave hitting us side on, but I can't really remember as I was asleep. All I remember was waking up in mid-air, then making a big "thud" against the door.' And at over 180 cm in his socks, Jeff is not what you'd call a 'slightly built' man.

For a bloke who enjoys a challenge and a bit of a tinker, it's hard to picture a job more suited to Jeff. Think of him as a kind of mad inventor working by himself for weeks on some crazy idea in a shed at the bottom of the garden. Only he's not quite by himself, the garden is the engineering workshop in Hobart, and the crazy idea has been brought to him by a respected scientist who has just scored some invaluable time onboard *Southern Surveyor* to carry out possibly the work of a lifetime. Jeff listens to what they want to achieve out there on the water, then goes ahead and adapts or invents something that will do it for them. 'Yes', he says, boiling it down to the basics, 'the scientists come to me with an idea, and we bang something together.' He's part of a team, of course, helped by mechanical and electronics engineers such as Matt Sherlock who construct, test and adapt, but it's a job that often entails making something from scratch. Like the EZ net.

'Marine scientists like to know what creatures are in different depths of water', says Jeff. 'We could just drive a net through and go down to 400 m and bring it back up again, and it'd be full of critters, but you won't know what depth they came from.' The EZ net goes a long way to solving this dilemma. 'We adapted a frame with ten or so nets on it that we engineered to open and close at different depths. We can bring it up

The EZ net is hauled back onboard the vessel after being deployed to collect plankton and small fish samples. It has ten separate nets that can be opened and closed remotely at specific depths from the operations room and it can be used down to 1000 m. Source: MNF.

say, 50 m at a time, close one net, bring it up another 50 m, then close that one, and so on. It's gives a far more comprehensive picture of what's there in the water column.'

Jeff's job usually involved taking something basic and adapting it into a tool of the trade. 'A lot of these things were prototypes', he says. 'You don't realise certain problems until you put them in the water. You build it up in the workshop, and it all looks fine, then when you get to sea you think, "Oh, that's not quite right." My job was to support these one-off, icon instruments I'd helped build.'

I asked Jeff if he thought the scientists appreciate what he does and the problems he faces. He laughs at this. 'You know, the best possible thing a scientist could see me doing was sleeping in a beanbag. That meant that everything was ticking along well and I was able to rest. If we're running around, we're holding up the ship and everyone's waiting.'

One member of the science community he remembers with something less than fondness is an American working as part of a Defence contract job undertaken several hundred nautical miles north of Darwin. Specialised equipment, anchored to the seafloor, was retrieved by means of an acoustic release, where a hydrophone is lowered into the water to emit a frequency that springs the anchoring mechanism. 'Before I could say "Stop",' says Jeff, 'this American had lowered the hydrophone over the side and it was sheared straight off by the propeller. He just held this shredded hydrophone cable in his hand saying, "Have you got any spare parts?"' Jeff was able to connect it to one of the ship's echo sounder transducers. 'It didn't work quite as well, but after a couple of hours we managed to get it to release and bring the equipment to the surface to be recovered.' Having saved this passenger a couple of million dollars worth of equipment, Jeff felt pretty certain he wouldn't be seeing a bill for a beer when they got to shore, but it was not to be. 'He didn't even say goodbye, let alone buy me a beer.' Not that he's bitter. 'Says more about him really, doesn't it?' he adds, laughing.

Three-quarters of Jeff's years at sea have each entailed at least 85 sailing days, and asking him whether he's ever toted up the hours, days or even number of his voyages simply makes him look exhausted. As a self-confessed 'late starter', he has a young family these days and so is cutting down on his sea time, as he misses his 'little people' when they're not around. I ask him what's changed in the years he's been at sea with the *Surveyor.*

'Safety', he says without hesitation. 'We could do whatever we liked back then, now safety is paramount. That's been a major cultural change. For example, we deploy instruments these days in much calmer waters than we would have twenty years ago.' And it's not just physical safety. The issues of mental health and conflict resolution are also taught and taken seriously. The science community using the *Surveyor* has evolved also. 'It could be pretty random back then', he tells me. 'Scientists could be poorly organised. They'd get onboard and think they could do whatever they liked – suddenly deciding to use a piece of equipment days sooner than they'd originally planned, so we had to rush to get it ready. It wouldn't happen so much these days.'

A voyage didn't have to be spectacular to be memorable, he tells me. Some simply involved ordinary days at sea with everyone working well and everything just coming together.

'A couple of years ago, we did a survey in the Great Australian Bight. It was a very ambitious twenty-one-day voyage with quite a lot of specialised hydrochemistry involved', he says. 'Matt and I built a thing called the ICP (integrated coring platform), a sediment corer which we could lower down to capture 300 mm of the seabed, the composition of which the scientists were very interested to know about. But it's not that easy to do when it's 3 km down.' One of the problems, he says, was knowing when the device had hit the seafloor, so cameras and fibre-optic cables were rigged up and attached. 'It let us see the bottom so we could bounce this thing along the bottom till it picked up enough sediment.'

It was a voyage involving improvisation and hard work on the run but in the end the scientists, who thought they'd have to drop some aspects of their ambitious schedule, had their expectations exceeded. The weather played its part too. 'We were out there for twenty-one days and got twenty-one days of perfect weather. There was a fantastic team feel about it all, and the work, from a conservation point of view,

The integrated coring platform (ICP) has six clear polycarbonate tubes which are pushed into soft sediment such as mud to collect samples to understand the seabed composition, from microbes through to small invertebrates. This data is especially useful for oil and gas exploration. The ICP can be fitted with other equipment including cameras and water sampling devices. Source: MNF/Mark Lewis.

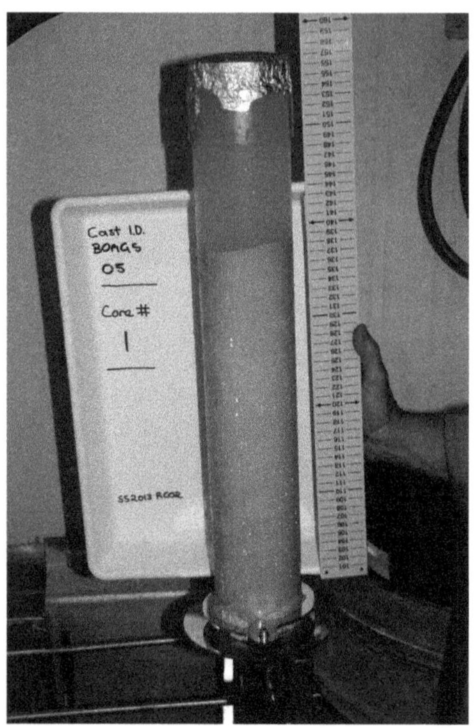

A core sample from the ICP is measured, then extracted from the tube and cut into sections for analysis. Different scientists require different sections of the core, for example the first 3 cm is used by biologists and chemists. Source: MNF/Mark Lewis.

is pretty important. You're discovering what's down there on the seafloor. You're pretty proud to see this thing going in the water, knowing that it answers an important science question and it's the only one in the world. Oh, and that you built it.'

At this point, a slow smile spreads across Jeff's face and he seems to focus on something in the middle distance. 'Some of what we do really is of global significance', he says. 'Like fish stock assessments, having some influence on a particular species not being overfished, or helping in its recovery. Not all of it is exciting. You might be going away for a month when all they're doing is bringing up water and analysing it, day after day and it's boring as. But a year later, they've made a discovery about an ocean current moving differently than expected. And you've played your small part in it. It's a great feeling to be a part of something like that.'

Lindsay Pender

Physicist/marine systems engineer

> *'The oceanographic information that came back from the SeaSoar was always amazing. It really showed you how dynamic the ocean can be.'*

To illustrate precisely what he means by 'ocean turbulence', Lindsay gets me to stick my spoon in my coffee and give it a swirl. 'What you're doing there is mixing up the milk and water', he says, watching it all closely, 'inducing a current and creating a shear – two bodies of liquid moving against each other.' These processes, he tells me, are also happening in the ocean, driven not by my spoon but by global ocean currents. 'It's one process that transfers heat from the upper ocean to the deeper parts and moves nutrients up to the shallower parts. At the interface of the two moving and stationary bodies of water, little eddies are set up and that's what we mean by ocean turbulence. It's the process that mixes the two bodies of water.' I swear I'll never look at my latte the same way again.

Lindsay Pender is what you call a practical man. As the Marine National Facility Technical Advisor, he's been with CSIRO a good many years, starting in 1985 when he was employed to work on a specific instrument which measured, yes, ocean turbulence. Going back, Victorian-born Lindsay completed a PhD at Monash University looking at the electrical properties of polymers, then postdoc work in Canada and later at the Australian National University where he examined interaction of high-energy ions with solid forms. Today he happily describes himself as a 'jack-of-all-trades'.

I gather from what he tells me that the instrument that brought him to CSIRO in the first place looked rather like a torpedo with probes containing something like gramophone needles sticking out the front to pick up the tiny variations in water turbulence as it was towed behind a ship. Lindsay was chiefly involved in the computing side and the design of the control systems of this complicated instrument, but as he describes its function and explains points such as how the 'angle of the combined vector relative to the axis of the airfoil is fairly small', I'm struggling to keep pace. But then he brings it back down to earth. 'We called it "Bunyip"', he says,

Mark Underwood and Lindsay Pender work on the SeaSoar as Iain Suthers watches. The SeaSoar was fitted with sensors to measure temperature, oxygen levels, salinity and turbidity. Source: MNF.

'which was quite appropriate because we never actually got it going properly – so it was all rather mythical!' Despite its ambitious description, the poor old Bunyip was abandoned after a few years, chiefly, says Lindsay with a laugh, 'because we didn't understand the data we were getting from it. It just didn't make any sense!'

But if the Bunyip proved a failure, another instrument, the English-manufactured SeaSoar, is a story of unqualified success, at least for a time. Officially a 'towed undulating vehicle' with a price tag of about $250 000, Lindsay and his team added a few modifications so that, he says, 'it didn't even look like a SeaSoar by the time we'd finished with it. We added larger wings to increase the depth range, we improved the telemetry so we could talk to it faster from the ship, and we improved its stability.' This new-look SeaSoar proved so effective that it became the workhorse for much of the oceanographic research undertaken by *Franklin* and *Southern Surveyor*. 'The oceanographic information that came back from the SeaSoar was always amazing', says Lindsay. 'It really showed you how dynamic the ocean can be.' At one stage, Lindsay visited the factory on the outskirts of London to suggest one or two of his improvements.

One adaptation, however, was not so successful. The story starts in the middle of a spectacular-sounding voyage to Australia's north. 'We were looking after the oceanographic side of a big international project researching "Hector"', says Lindsay.

'Hector' is the name of a dramatic thunderstorm system which develops daily in the wet season over the Tiwi Islands and drifts towards Darwin. 'We were looking at the air–sea interface', he says. 'It's a really impressive storm – lots of lightning and thunder – and it was a pretty intense operation with lots of other countries involved. It was quite a big deal. There were atmospheric researchers, and some scientists were even in aeroplanes.'

One stormy afternoon returning from a quick run into Darwin, they decided to deploy the SeaSoar, towing it behind the *Surveyor* in choppy shallow water which turned out to be too rough for the instrument. 'The tailplane just broke off', says Lindsay. 'In hindsight we shouldn't have been operating in those conditions, so if anyone was at fault it was me.' The tailplane also housed the SeaSoar's stabilisation mechanism and, without it, it was essentially uncontrollable. Necessity being the mother of invention, a quick scrounge around the bowels of the ship proffered some sheet metal 'and other bits and pieces' which, with the aid of a hacksaw and a welder, were bolted onto the SeaSoar in the *Surveyor*'s workshop to allow it to continue its work. 'We got it going again but it was never as stable', says Lindsay.

Even in its patched-up capacity, the SeaSoar continued to do its job. But with only an improvised stabilisation system, it seemed to fly 'like a bit of a dog'. 'Every now and again it would suddenly do a full 360° flip.' It was not realised at the time, but this strain began to impact on the cable termination, the point where it attached to the tow bridle. 'Because it was more or less working, and because the scientists wanted to keep using it', he says, 'we were unaware that with every one of these turns, metal fatigue was accumulating on the bridle.' Two years later, one night in the Pacific out towards New Caledonia, the inevitable occurred and the SeaSoar, all quarter-of-a-million dollars of it, parted ways with its cable and 'soared' straight to the bottom of the ocean where it presumably remains to this day.

Others I have spoken to who were aboard on this memorable night remember the slightly numb feeling as the readings from the SeaSoar suddenly went blank. 'We were putting 400 volts down that cable to power it', says Lindsay, still in a slightly subdued voice. 'The first thing that would happen was that everything would just short out. You'd soon know that something serious had gone wrong.' Such an ignominious end for the noble SeaSoar.

When I ask how many voyages Lindsay undertook on *Southern Surveyor*, a sudden look of exhaustion comes over his face. He hesitates to put a number on it, but estimates an average of three per year from when he first went to sea in 2002. Onboard ship, Lindsay tended to divide his time pretty evenly. 'If I wasn't sleeping, I was working', he says. Sure, one could watch the waves and the seabirds but in the main Lindsay saw his ship time as an opportunity to throw himself into the part of his job he really loved: developing software and engineering solutions, 'as distinct from a lot of the bureaucracy that goes with it', he says. And these were skills in great

Southern Surveyor deployed the tsunami early warning system and the Southern Ocean Flux Moorings. She mapped the seafloor to better understand the ecology and geology of Australia's marine estate and to discover the dynamics of Pacific tectonic plates. She also enabled a better understanding of the western rock lobster's life cycle, plankton, food chains and ocean currents in the seas around Australia. Source: MNF.

demand onboard *Southern Surveyor* as scientists sometimes struggled with data they were receiving and the means to interpret it. 'There were many situations where you'd be working with a scientist interested in some particular field who found they didn't have the capabilities to interpret and develop the data they were receiving. So I spent a lot of time developing software to enable them to do that sort of thing', he says.

Another problem that presented itself to Lindsay concerned Tom Trull's deep sea pulse moorings, which simply wouldn't stay in the same place. Attached to a cable anchored to the sea bottom, this instrument was designed to take water samples at the ideal depth of 20 m below the surface. The trouble is that tides and currents intervene and drag the mooring much deeper, depending on their strength. 'With a mooring anchored in about 4000 m', explains Lindsay, 'it wouldn't be abnormal to have it pulled over 200–300 m.' The obvious solution was to suspend it from a floating buoy. But the constant jolting of wave action would eventually most likely jar the instrument to pieces and break the buoy free.

The solution – hit on by Lindsay – was disarmingly simple. 'I used bungee rope', he says. 'We had a system where it floated on the surface but was tethered via a big length of bungee rope. It strings out, but doesn't get pulled down.' Thirty metres of bungee rope, absorbing the action of the waves, did the trick and kept the scientists, the instrument and Lindsay happy. But the development of this system wasn't without teething troubles. 'We did lose a buoy', he says. 'It eventually washed up in New Zealand and ended up, I believe, as a piece of garden furniture!'

One of Lindsay's most enduring memories of his years at sea occurred on the *Franklin*, but is worth mentioning simply to further illustrate the potential hazards of undertaking marine research at the bottom of the globe. 'It was quite a way down south', he says. 'We had this really large swell coming through, followed by an incredibly intense front. And then the wind suddenly swung around 90° to the sea.' A 'perfect storm' situation such as this – where the wind, sea and swell are cutting across each other at crazy, violent angles – is what every mariner dreads. 'The Master was very, very concerned about how the ship was going to behave in this situation', says Lindsay. 'What was he to do? Where do you point the ship? Beam to the wind or the swell? Which do you choose to get blown over by?' It was serious enough to warrant the entire crew being made to don their full-length emergency escape gear, and gather in the ops room to wait out the gale. 'The front only lasted a couple of hours', says Lindsay, 'but it was incredibly intense. On the map you could see the isobars stacked up solid black against one another.' At the time he didn't feel afraid, nor did he lose confidence in the ship's well-rehearsed systems. 'Obviously the Master was pretty concerned about the whole thing.'

The storm obviously left a lasting impression and it's a sobering image to leave me with, although not entirely. Much of Lindsay's sailing on *Surveyor* took him to the tropics, to New Guinea and the Bismarck Sea. There, he says, it was a different picture indeed, and it's the birds which leave a deep impression. 'Up there it can be idyllic', he says. 'You can be steaming along in straight line on a glassy sea and see the wake reaching back dead straight behind you. Then, in the southern waters, you had the seabirds. I never get sick of watching the albatross. Just their ability to never beat a wing, to follow the updraught of the waves. Glorious birds.'

Mark Lewis

Equipment specialist

> *'She was a great ship. I've been down to Macquarie Island on her in 16 m swells, and she was fine.'*

Mark takes me straight to the bottom – the bottom of the ocean that is, courtesy of an extremely rugged piece of equipment which he sounds just a little proud of, called 'Sherman'. 'Built like a tank', he assures me. Sherman is what's called a hard bottom sampler, in essence an underwater sled, constructed from over a ton of steel and designed to scoop up seafloor biological samples with its gaping, 1 m wide mouth. 'If the undersea terrain was smooth you'd use a beam trawl', he explains, 'a 4 m wide pole with a fine mesh at the back'. But when the going got tough and it was the seamounts and hard reefs the scientists wanted to sample, you'd call for Sherman. It was lowered gently off the *Surveyor*'s stern deck to the bottom, then towed – well, scraped really – along a transect for no more than ten minutes at a time. 'That's how long it usually took to fill up', says Mark. Then it was hauled back up for the scientists to open it, get their samples and go to work. I ask what would happen if it hit a rock. 'It went "bang"', he says.

In fact, Sherman stopped the *Surveyor* in its tracks on three occasions. 'You're steaming along', says Mark, 'and suddenly she's not going so fast, then she gets slower, then she pulls up entirely.' To me it sounds incredible that the *Surveyor*'s 1100 tons could become snagged by a device designed to collect seaweed, but Mark assures me it's true. 'Then we'd have to pull backwards onto it and try to get off', he says. The scenario reminds me of my brief fishing days, as a kid, spent mainly manoeuvring around rocks and river-banks, rod in hand, attempting to locate the usually impossible angle that would free my tangled line.

Mark loved working onboard *Southern Surveyor* and speaks about her, and everything else for that matter, with enthusiasm. He's just that kind of guy. 'She was a great ship', he says. 'I've been down to Macquarie Island on her in 16 m swells, and she was fine.' Such seas sound utterly terrifying, I tell him, but he assures me he felt quite safe and that everyone 'just locked all the doors and kept going'. Mark had

Mark Lewis helps to prepare a Sherman sled for deployment on a transit voyage as part of the MNF's Next Wave program, which was designed to give students a taste of what it's like to live and work on an ocean-going research vessel. Source: MNF/Matt Sherlock.

suffered seasickness on smaller vessels, but never on *Surveyor*. 'There was something about the motion of her I very much liked', he says.

She was also, he tells me, an excellent work platform. 'We did a lot of fishing and launching and towing from the stern, and with her A-frame and stern ramp all set up from her days as a trawler, it gave us a great deal of flexibility.' And as an equipment specialist, Mark needed to be flexible. On a typical voyage – many of his involved habitat mapping – the scientists and hydrographers would use the swath mapper to survey an area of interest. As that information was being processed, Mark and his team would set up an underwater camera system and launch it off the stern. 'After we'd done a couple of camera transects down to a maximum of 670 m, the scientists would take a look at the video images and decide where they wanted to take their samples.'

Rocks and shelves had to be taken into account and avoided, but they weren't the only hazard. 'We'd often see on the screen bits of fishing gear and nets still caught on the bottom, and we'd have to quickly pull up because if the camera gear got caught in it, you could lose it', he says. 'It's something you really had to watch out for.'

On one occasion Sherman sampled something rather extraordinary. 'Down south one time there were some cod which had burrowed into the coral', Mark says, 'but Sherman just scooped up the lot, coral, cod, everything.' Out of the five fish caught in this unlikely manner, three were new species and the others were new records for Australia. 'The onboard fish taxonomists were delighted and came up to us very excited, "These are really good!"' All agreed it was quite the catch of the day.

Mark Lewis preparing a suite of sensors, which look at the chemical composition of the ocean and are attached to the integrated coring platform. Time onboard the MNF vessel is valuable and shifts run twelve hours on and twelve hours off, with experiments and data collection happening throughout the night on most voyages. Source: MNF/Matt Sherlock.

The best description of his job was provided by Mark's colleague, electronics specialist Jeff Cordell, who dubbed him the 'Minister for Heavy Things'. Between the two of them, they seemed to have worked out the demarcation. 'If it had batteries it was his', says Mark, 'if it had barnacles it was mine.' An apt description, as Mark was responsible for the preparation and deployment of a serious array of heavy-duty equipment: camera systems, sleds, nets, trawl gear and of course beasts such as Sherman, for much of which he'd overseen the construction at the CSIRO Hobart workshops.

Then there were highly specialised gadgets such as the coring systems, those arrays of what sound like plumbing pipes used by scientists such as Patrick De Deckker, attached to a heavy steel frame and dropped into the seabed to capture the sand, mud or whatever makes up the surface. But with sea time precious and big chunks of it being taken to simply lower and raise equipment thousands of metres to and from the seabed, Mark needed to think laterally. 'We had to make the most of the transit time to the bottom, so inside the coring system we built a framework', he says. 'Onto this frame we'd mount three cameras so we could see the bottom, then transducers to get acoustic information as it was going down through the water column.' Added to this array would be a CTD to read water quality, and hydrocarbon 'sniffing' equipment to look at methane and hydrocarbons. 'Then we'd lower the whole thing. This one bit of kit gave us very intensive sampling.'

Lowering this leviathan sometimes several kilometres to hit a precise spot on the ocean floor sounds difficult enough, but what if you needed to do it twice? This is where one of Mark's favourite features of *Southern Surveyor* made life easier. 'The big advantage of that vessel was her dynamic positioning system', he explains. 'One occasion, in about 2000 m of water, we landed the corer but it had misfired, and the doors had shut prematurely so we couldn't get the sample.' There was nothing for it but to haul it back up, reset and start again.

'Keep position please', was the request to the bridge from the operations room, to try to find that exact point. 'We reset it all and sent it back down, and in the cameras, you could actually see the same spot on the bottom where we'd landed before', says Mark. 'That was how far she'd moved – not at all.' Mark reckons it was the equivalent of standing on top of Mount Wellington with a fishing rod, trying to cast a fly onto a street intersection way below. Twice.

The *Surveyor* was also, Mark tells me, extremely manoeuvrable. Sturdy, I had been told many times, as well as reliable, but her handling ability was also impressive. 'Once we had a camera system caught on a nasty little cliff we'd run into somewhere off WA', he tells me. 'It was at about 600 m depth and we couldn't get it off by going forward, so we were able to reverse back and used the thrusters, manoeuvring her around and pulling it off the other way.' You could apparently get a sense of *Surveyor*'s handling skills watching her come into port. Instead of requiring the pushing and pulling of tugs, *Surveyor* 'would come in, do a handbrake turn and park at the wharf. She could go sideways into 25 knots of wind.'

Other aspects of *Surveyor*, however, were not quite as user-friendly. The ship's surface towing platform, from which much of the equipment was deployed, was somewhat inconveniently placed just aft of not only the galley's maserator chute, but also the ship's sewage outlet. This could make life particularly unpleasant when deploying instruments such as the fine mesh manta net – so named for its two gaping wings at the front resembling the great ray's mouth, and used for gathering surface debris. It was not built to collect the ship's, not to mention the crew's, waste! Luckily there was a system in place to prevent such an occurrence. 'There was a big button down in the operations room called the "poo button"', explains Mark, 'which you had to hit if you were deploying a surface net'. Neither of us quite manages to keep a straight face as he tells me this. 'I mean it was all treated', he continues, 'but it still wasn't pleasant to sort through!'

Mark was called on to build all sorts of pieces of equipment for all sorts of areas of research, from exploring seamounts to a cage especially built to house oddities such as the gulper shark. I'd never heard of this apparently endangered and (if the pictures are anything to go by) unfortunate-looking fish which Mark describes as being a little like the 'runt of the shark world'.

'They're a slow, low-metabolic rate shark, not very active predators, a bit lazy really', he says. A little research tells me these odd-looking creatures inhabit deep murky water and pretty much keep to themselves, rarely going after anything but happy to take a chomp at something if it crosses their path. Hence, their numbers have been severely affected with the advent of long-line fishing practices.

On one trip to the Great Australian Bight, the brief was to catch some examples of the gulper shark, bring them to the surface, fit them with an acoustic tag and release them. The scientists were unsure of the effects on the sharks of dragging them up from their 400 m deep habitat and lowering them back down. Mark came up with a solution. 'We built a trap with an automatic door on it – a 2 × 1 × 1.5 m box of mesh with a bait to attract them and a video camera, so we could watch them as they descended again to the bottom.' It turns out they were fine. 'The door opened', he says, 'and they just wandered back out again.' Then an array of listening stations on anchored moorings would come into the picture, picking up the signals from the sharks' tags, tracking them individually as they swam past. 'Oh, look, 3461's just gone by', is how Mark describes the technique, which gave scientists a previously unseen picture of the habits of this reclusive animal.

Other sharks weren't so benign. 'They used to come in and bite the cable', he says. 'You'd haul it in and there'd be bite marks through the outer, and nearly through the inner cable too. They'd just take a chomp out to see what it was.' Some moorings, he reckons, were lost this way, a fact discovered when their retrieval was attempted. 'We'd locate the acoustic signal, but actually their float had been bitten away, releasing the data recorder which floated away to who knows where?' Some managed to find their way back to CSIRO, courtesy of curious and conscientious fishermen who, after finding them, called the number that was always prominent. 'We always let them keep the cable and the float', says Mark, 'all we wanted was the little grey box with the flashing LED. We even paid for the postage!'

One of his best memories, he tells me, was a voyage from Fremantle to Albany then up as far as Broome, carrying out some intense habitat mapping. 'What was really fun was getting the data and seeing the response from the sponge researcher from the Western Australian Museum. She was over the moon, never having seen sponges this good, or this big or this colour. She had to hire a freezer container because she had so much material she couldn't put it at the museum. That was very satisfying.'

Mark reckons he averaged about eighty sailing days on *Surveyor* for about ten years, but has never properly totted it up. 'I have a four-drawer filing cabinet in my office with three of them taken up with voyage notes if you want to see them', he suggests, but judging by the enthusiasm with which he remembers his time onboard I suspect he hasn't forgotten a moment of it.

Tara Martin

Geophysicist

> *'Sometimes a lot of the science of what we do seems like tiny little grains, but those grains add up.'*

Geophysicist Tara Martin came to *Southern Surveyor* relatively late, getting to know the ship only during her last couple of years with the MNF, but in that short time she managed to cram a virtual career's worth of interesting work. Out of the four hundred-odd sailing days of *Surveyor*'s last two seasons, Tara was onboard for 231 of them. 'I got quite fond of the old girl', she says. 'In the end the crew were starting to have bets as to who was spending more time aboard, them or me.' Tara, I soon discover, has the gift of making everything she says sound completely fascinating. Whenever she spoke, I found myself drawn to the edge of my seat, hanging off every syllable of her delicate, quietly excited voice. I ask if she remembers her first voyage with *Surveyor*. 'Yes', she says effortlessly, 'voyage SS 2011 CO1 – a Bureau of Meteorology research charter, laying out and retrieving tsunami monitoring buoys.' She proceeds to explain how these buoys work, about the strain gauge which detects larger than usual waves and how, having been placed in response to the 2004 Boxing Day tsunami, they are now strung along the east coast of Australia far out at sea, providing a vital warning to coastal populations.

At one stage she found herself scanning the ocean, trying to spot a floating buoy which had just been released from its mooring on the seabed. 'As I was straining to spot this small pale-grey object in an area anything up to two square miles, on a grey ocean in thick fog, it suddenly struck me, "Who on earth thought it was a good idea painting these things grey?"' We agreed that red would have been a far more sensible colour.

Tara had other interesting facts about these ocean buoys, such as how their anchors are often old train wheels. 'Some of the larger buoys can be on cables stretching 5 km long, and very heavy', she says.

Tara's job on the first of her many voyages was bathymetric surveying, and she offers some interesting insights into the function of the *Surveyor*'s swath mapping system. 'It's really a depth sounder on steroids', she says. 'The main reason it sits in its gondola beneath the hull is to avoid the ambient vibrations from the engines and other parts of the ship.' (I was told by just about everybody who worked on her that *Surveyor* was a noisy lady!) 'We create vibrations in the water in exactly the same way my larynx works enabling you hear my voice now', she says. A series of piezo ceramic plates are vibrated at a certain frequency, sending a sound pulse through the ocean. 'And from there it's a perfectly simple equation', says Tara. 'It goes to the bottom, it bounces back and we count how long it takes.'

The swath is not one frequency beam but hundreds, spreading out like a fan across the seafloor and returning a picture which varies according to the depth of the water. *Southern Surveyor*'s is a midwater system, Tara tells me, able to receive an accurate picture five times wider than the water's depth. 'But in more than about 2000 m, we'd have to draw that beam tighter and tighter and get a much narrower resolution.'

For at least twelve hours a day, behind a rack of computers in a small windowless L-shaped corner of the *Surveyor*'s operations room, Tara would sit glued to her screen, watching the data come in. She makes what to many would seem a tedious task sound almost inspiring. 'I read those images in the same way as you would read a book, or someone's face', she says. 'I forget that other people don't see what I see.' It's more than simply the depth of the water that's being gauged. To an experienced reader such as Tara, the swath mapper can give a highly detailed picture of the seafloor, determining whether it's sand or mud, coarse or fine sand, smooth rock or rough. I ask whether a single boulder can be picked out down there in the cold murky depths. 'Oh yes!' she says brightly. 'Boulders just leap out at you!'

With a machine that can deliver details such as this, Tara almost guesses my next line of enquiry – shipwrecks. 'We've certainly done some interesting ones', she says, and gives me the story of the MV *Limerick*. Built in Belfast in 1925, the 9000 ton freighter *Limerick* was a fast ship for her day and in April 1943 was in a convoy of five vessels steaming from Sydney to Brisbane with a load of iron ore in her holds. Just after 1am, ~12 miles off the New South Wales coast, a marauding Japanese submarine slammed a torpedo into her port side. She caught fire immediately, sinking just after 6am and taking two crewmen with her.

For seventy years, her exact whereabouts remained a mystery. Then in 2012, a NSW couple decided to show a friend their favourite fishing spot which they'd stumbled upon some years before and had previously kept to themselves. The friend, possessing perhaps a slightly different sense of perspective, put it to them straight: 'This spot of yours. It's in a dead flat part of the seafloor and sits 10 m above the seabed and is about 100 m long. Don't you think you should tell someone about it?' The couple, their conscience perhaps stirred, saw his point and contacted the NSW Department of Environment and Heritage. A police launch was sent out to investigate, a sounding was taken, and Tara Martin was hunting for a World War II shipwreck.

'It just so happened we were already in the area doing a voyage with Tom Hubble from the University of Sydney, looking at potential landslides on the continental slope, when the Department of Heritage and Environment asked us to go and take a look at it', she says. Tom graciously agreed to surrender some of his precious research time to allow the *Surveyor* to make this unscheduled diversion. 'We did a loop around the wreck area and were able to determine that it was the right size and the right shape for the *Limerick*, and I'm pretty sure I could see the torpedo hole.' Having always had a fascination with shipwrecks, this detail elicited from me a slight gasp.

Against the prevailing views of some of the experts onboard, Tara started to develop a strong suspicion that the vessel had come to rest in an inverted position. 'Oh no, it's very unlikely the ship would have turned turtle in the water', they assured her. But whichever way the *Limerick* was resting, they were all pretty confident she had been found. Later, a diver undertook the challenging 100 m descent to confirm she was indeed the *Limerick*, that one of her two propellers was missing, that a massive torpedo hole was evident in her port side and that she was indeed resting upside

Multi-beam sonar image showing the MV *Limerick*. Tom Hubble from the University of Sydney was allocated time onboard *Southern Surveyor* to conduct geological research along the continental slope and shelf between Yamba and Fraser Island in early 2013. The NSW Department of Environment and Heritage asked if he would give up some of his research time to locate the shipwreck of MV *Limerick* and he was pleased to assist, because finding the shipwreck was in the national interest. Source: MNF.

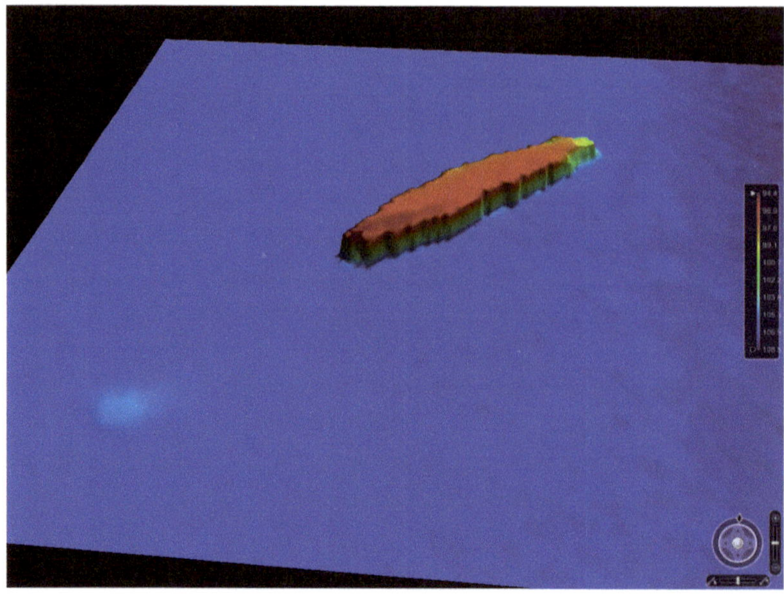

MV *Limerick* was a 140 m merchant ship built in 1925 in the UK and, at the time of sinking, it belonged to the Union Steam Ship Co. in New Zealand. On its way from Sydney to Brisbane, the vessel was torpedoed at 1am on 26 April 1943 by the Japanese submarine I-177. There were seventy-two crew onboard when it sank and two went down with the ship. Source: MNF.

down. 'Our theory is that her cargo of ore became wet and shifted', says Tara. The *Limerick* is now a protected historic shipwreck, and something tells me it's one of Tara's proudest moments aboard *Surveyor*.

The voyage was also memorable, Tara tells me, for another reason, namely the case of *Southern Surveyor*'s only known stowaway. It's hard to tell if she's being serious. 'One January night', she begins as I lean forward like a kid about to hear a bedtime story, 'our voyage manager wandered onto the deck to take in the air and enjoy an apple. As he's looking over the side he feels something lick his leg!' (I begin to wonder where we're heading with this.) 'It was Patrick, the seagoing possum!' she says, emitting a sudden laugh. It appears that while the *Surveyor* had been laid up in a Brisbane shipyard one Christmas, the adventurous marsupial decided to quietly join the crew. 'We think that's where Patrick joined us', she says. I'm curious to know the voyage manager's reaction upon discovering a possum onboard and staring straight at him. 'He gave him the apple', says Tara.

While a stray possum might not be a catastrophe, when it's discovered on day one of a sea voyage lasting several weeks, it's bound to affect the nature of the trip. 'It immediately raised both the hunting and nurturing instincts of the whole crew', says Tara, as if describing some sociological experiment. A hunt for the animal ensued the next day and it was found sleeping, somewhat inconveniently, in a winch drum. 'We

Patrick the stowaway possum was found a day into an eighteen-day voyage that departed in January 2013 from Brisbane. Source: MNF/Karl Forcey.

caught it, and the crew made a cage for it out of five milk crates tied end to end with cable ties, the bottom cut out of three of them', says Tara.

For a time all went well for the possum, now dubbed 'Patrick' and relocated, cage and all, to a fish laboratory next to a scupper for fresh air, and fed three times a day. 'That was honestly one well-fed possum', she says. Then, as the tail end of tropical cyclone Oswald struck northern New South Wales with force ten gales and high seas, the *Surveyor* was forced to heave-to for twenty-four hours and ride it out off Noosa. 'At some stage during this I walked into the fish lab and realised the waves were so high they were washing in through the scupper', says Tara. Alarmed, Tara made her way to the bridge where the Third Officer was on watch, explaining to him that Patrick might be in danger of being inundated. The officer reached for his radio. 'Guys, guys, we've got to rescue the possum!' he decreed.

'We found Patrick, drenched but otherwise unharmed and relocated him to the CTD area', she says. From there, his adventures continued. One time he chewed through his cage's cable ties and found his way back to the winch drum, then he vanished again for days on end after his intact cage was discovered, mysteriously empty. That part remains a puzzle still.

Eventually, the *Surveyor* re-berthed in Brisbane, not too far from where it was assumed Patrick had come aboard. At dusk, he was escorted in-cage to the bottom of a large tree and released, thus ending the Patrick saga. 'Although next morning there were rumours he was spotted heading back towards the ship', says Tara. Perhaps the Patrick story has yet to run its course.

As Tara talks about her work, I realise that it requires neither shipwrecks nor wayward possums to make it sound interesting. When engaged in seafloor mapping, Tara regularly encounters parts of the world unseen by human eyes. I ask her what goes through her mind when she sees the images and the data. What's it like down there? I'm curious to know. Are there hills, valleys? 'On a scale you can't possibly imagine', she says quietly. 'More people have been to the Moon than have been to the bottom of our oceans.' Indeed, her descriptions make it sound like some other world. 'The pressures are immense. You gain an atmosphere every 10 m. There are vast canyons, immense volcanoes, mountains towering 3000, 4000, 5000 m above the seafloor which don't touch the surface. You lose a sense of scale down there very easily.'

Having believed that modern technology and satellites had long since accounted for every nook and cranny of the planet, I'm a little shocked to learn that barely 25% of Australia's oceanic area has been mapped, with scientists having only a rough idea of the rest of it. 'Satellite information can be way, way out of line, and for that matter, so can the charts', she says, and relates a story which strictly speaking predates her time with the *Surveyor* but illustrates her point as well as the genuine pioneering nature of her work. 'I did a voyage with the British Antarctic Survey around the South Sandwich Islands between South America and the tip of Antarctica', she says. It's the sort of place that sends a chill running through you just looking at it on a map, if you can find it. When surveying this vast undersea volcanic chain, Tara one evening walked into the operations room to see the Chief Scientist sweating and visibly shaken. 'I nearly just killed us', he said, looking up to an astonished Tara. 'We've just cleared a seamount by 17 m', he said, mopping his brow. 'Now, the chart', Tara explains, 'told us this was 220 m below sea level, and given that the ship had a draft of 6.5 m it was pretty close.' Days from help in one of the most inhospitable spots on the planet, a collision with the solid dolerite tip of this dormant undersea volcano would have been catastrophic. But thanks to the work of people like Tara, the charts – corrected – will hopefully prevent others from courting such a disaster.

It's not simply in terms of navigation and correct charts that the importance of Tara's work can be measured. 'We want to understand what's down there for fisheries management, or where we might find mineral resources', she says. Hence, no part of *Southern Surveyor*'s travel itinerary is wasted, even in transit. Where possible, the vessel's path will follow an unexplored line so as to incrementally build up the data, gradually filling in the vast blank canvas of the seafloor.

Sometimes that canvas can be a little too blank, particularly when deploying valuable pieces of equipment, as on one of Patrick De Deckker's School of Earth Sciences voyages. As ocean sediment samples were being taken with one of the professor's large heavy corers, Tara noticed her depth reading varying wildly as it was being lowered, jumping back and forth between 100, 150, 180 m. 'I said to myself, "What's going on?"' Then it hit her, and she leaped onto the radio to the bridge. 'Stop!' she implored, just in time to prevent significant damage to the unique and expensive corer.

'I realised they were lowering it down the vertical escarpment of an undersea cliff', she says. The movement of the *Surveyor* would have swung it violently against the rock face like a wrecking ball, undoubtedly causing serious damage. 'Could we possibly move over 20 m or so please?' she asked. Patrick was most grateful, and his valuable work was able to continue.

Of the many images of the submarine topography she has given me, I ask Tara what stands out particularly. She thinks for a moment before describing what lies offshore near where she grew up on the south-west Victorian coast. 'Out into the Great Australian Bight near the South Australian side, there's a series of absolutely round sink-holes in the limestone, perfect, as if someone's pressed a coin into it.' It's a phenomenon due apparently to the area's undersea limestone caves which dissolve gradually, collapsing inwards to form remarkable shapes and spaces. 'It's just spectacular, stunning, in fact. Some of them are 100 m deep with beautiful vertical sides', she says.

These days, Tara describes herself as a 'reformed research scientist'. She has for some time been primarily involved in technical support, helping prepare the platform for the new *Investigator*. 'I believe that for science to work well, it needs a good platform to work from', she says. Not that she's planning to retire from research: Tara's already counting on some extended time at sea to continue, as she puts it, 'her little corner of science'.

'Sometimes a lot of the science of what we do seems like tiny little grains, but those grains add up. I look back and I think that a lot of the work I've done is absolutely insignificant, but when I think of some of the projects I've been involved with in a small way, I just think, "Wow".'

THE BIOLOGISTS

Chris Wilcox

Marine ecologist

'There's a lot of plastic out there.'

I try not to look blank when Chris Wilcox gives me his job description, though I probably give it away when I ask him to repeat it, a little slower. 'Marine ecologist with a strong mathematical bent, and a special interest in the dynamics of natural predator/prey systems', he tells me. But the more Chris speaks about his studies, the more intrigued – indeed alarmed – I become by what he's discovered in his work on *Southern Surveyor* about the state of our oceans.

Chris's interest in the sea has a long pedigree. 'I've worked on everything from how swordfish and tuna move across the Pacific to detecting lawbreaking by fishermen', he says. Initially, however, like several the scientists I spoke to, Chris didn't work anywhere near the ocean, but in the desert. Originally from California, he somehow found his way to the Australian Outback via the University of Queensland, specialising in the effects on desert springs in the Great Artesian Basin of groundwater withdrawal for mining. At one point, he underwent a dramatic change of priorities. 'I felt I needed to realign my love of surfing and living on the coast with my job', he says wryly.

Although employed by CSIRO, Chris was subject to the same selection process as any scientist requesting use of the MNF. The strength of his work saw him awarded three voyages, all part of a major risk project in an area that seems, thankfully, to be garnering more and more public interest – the impact of plastics on ocean wildlife. 'It's pretty simple', he says. 'You look at where the wildlife is, then you look at where the plastic is and you overlay the two.' From this fundamental information, Chris assesses how much plastic is being encountered by which species, the impact in terms of entanglement and ingestion, then the overall impact on the species themselves. How this was actually done sounds ingeniously simple. Right around the Australian continent, shore surveys were carried out along beaches and coastlines every 100 km in conjunction with offshore surveys by *Southern Surveyor*, which took samples at key locations. 'Then we tried to put the two together to see what the relationship is. If the

Debris littered across a beach on Christmas Island. Source: Chris Wilcox.

onshore survey is a good predictor of what's offshore, we use an oceanographic model to forecast what ends up in the ocean', he explains.

To realise this grand undertaking, Chris and his team mobilised the powerful yet strangely untapped resource of … small children. As part of a citizen science program, about 200 delighted primary schoolers and some of their teachers combed beaches around the country collecting coastal data over a series of days that must surely have beaten anything that was happening in the classroom.

Out on *Surveyor*, a super-fine 300 micron manta net was towed just off the bow and ahead of the wake, skimming the surface and trapping any particles, large and small, that came into its maw. 'Lots of bits of plastic, most small hard pieces, much of it broken down is what we found', says Chris.

A book I recently read, which discusses the extent of plastic pollution in northern hemisphere oceans, prompts me to seek Chris's assessment of our own long and famous coastline. What he tells me is not particularly encouraging. 'We found the density of plastic around Australia in the marine environment got as high as 43 000 items per kilometre', he says, a figure which astounds me. And this, I am reminded, is what is being found out at sea, well over the horizon. 'That of course is only the relatively small amount of plastic that actually floats. The rest is most likely forming

The manta net has two wings that help it glide along the surface of the ocean. It has a fine gauge and a collection container at the end of the net. Source: MNF.

plumes on the bottom, particularly near our cities', he says. 'Yes, there's a lot of plastic out there.' He describes it as human society 'leaking' into the natural environment of the oceans.

The effect of this on marine animals is a large part of Chris's brief, and many a necropsy was carried out on the gizzards of seabirds and turtles found on the beaches around Australia to determine not just ingestion, but actual absorption of plastics into the animals' bodies. 'Plastics have a binding agent which keeps the fibres together and those leach out when they go through an animal's digestive system and get caught in their fat', he explains.

The worst culprit? The one item more responsible for clogging the ocean environment than any other? He doesn't hesitate for a moment. 'Drink bottles.' Chris and his team discovered that a whopping 40% of items found in coastal clean-ups were plastic beverage containers that we let slip from our hand the moment we have consumed the half-litre or so of flavoured sugary water within. And it's only going to get worse. 'The doubling time for plastic at the moment is eleven years', he tells me. 'Meaning that between 2014 and 2025, the world will make as much plastic as it made since we started making it fifty years ago. Then, between 2025 and 2036 we will again make as much as in that entire preceding time.' What about so-called biodegradable plastics? 'Not really a believer in them. Sorry', he says. 'It really just means it breaks down into smaller pieces.' Chris reminds me of the hard truth that an

WHAT DO PLASTICS IN THE SEA DO TO MARINE ANIMALS?

The CSIRO survey of the Australian coastline published in 2014 found that three-quarters of coastal rubbish is plastic, and that plastic in our nearby ocean ranges in density from a few thousand pieces of plastic to more than 40 000 pieces per square kilometre.

Globally, between 6350 and 245 000 t of plastic waste is estimated to float on the ocean's surface. Some plastic ends up in marine sediments where it may be ingested by bottom-dwelling creatures and filter-feeders. The rest is distributed through the water column, ends up on our shorelines or maybe in the guts of marine wildlife such as turtles, whales, dolphins, seabirds and fish.

Rubbish impacts wildlife directly through entanglement and ingestion and indirectly through chemical effects. Seabirds, turtles, whales, dolphins, dugongs, fish, crabs, crocodiles and numerous other species are killed and maimed through entanglement. Microplastics can be mistakenly eaten by a range of marine species. For example, turtles can mistake floating soft clear plastic for jellyfish, and globally about one-third of all turtles are believed to have eaten plastic in some form. Birds eat everything from balloon shreds to glow sticks, industrial plastic pellets, hard bits of plastic, foam, metal hooks and fishing line. Around the world, nearly half of all seabird species are likely to ingest debris. CSIRO researchers and colleagues found that 43% of short-tailed shearwaters have plastic in their gut, and there is evidence of the plasticising chemical Bisphenol A (BPA) in the oil glands of seabirds.

A global estimate of plastic waste in our oceans published in 2015 estimated that 8 million tonnes of plastic waste enter our oceans annually. That's the equivalent of 16 shopping bags full of plastic being placed on every metre of coastline for each of the 192 countries that border the Atlantic, Pacific and Indian oceans, and the Mediterranean and Black seas that were included in the study. Most of the worst plastic-polluting countries are middle- or low-income nations that are experiencing fast economic growth but are yet to implement the waste management infrastructure needed to handle the consequences. The research paves the way to improve global solid waste management and reduce plastic in the waste stream. By garnering the information needed to identify sources and hotspots of debris, we can better develop effective solutions to tackle marine debris.

average plastic milk bottle can take up to 500 years to degrade. 'We can't even think in that sort of time scale, and yet that's how long these things can last.'

If all this is making me thoroughly depressed, I'm heartened to learn that Chris remains something of the optimist. 'The thing is', he says, 'it's a solvable problem. Every one of those bottles was at some stage being held by someone before it was disposed of. You just have to change the incentive to dispose of it responsibly instead of just dropping it.' For this reason, Chris has long offered support for a container deposit scheme (currently in operation in just one Australian state, South Australia) and he has the research to back it. 'South Australia has one-quarter the amount of beverage containers in its waste stream as the other states', he says. 'When you think that simply having a container deposit scheme in place can cut the amount of plastic bottles by three-quarters, it's in some ways pretty satisfying. Of course though, you have to actually have them in place.' His optimism, I fear, far outstrips my own.

Forty-five councils around Australia were approached during the survey. It was found that those that had strict dumping and pollution regulations, and enforced them, had a significant reduction in plastic waste off their shores.

What was life like on the *Surveyor* while this work was being undertaken? 'We took three samples a day – three fifteen-minute tows – at 6am, noon and around dinner time', he says. The net was attached to a boom extending from the bow and towed at three knots before being winched aboard and its contents examined. 'We'd spend a few hours in the lab sorting through tubs of algae', says Chris. 'The plastic mixed up in it ranged from food wrappers to items that were just a couple of millimetres across.' Coming back from Fiji one time, when the seas were running at 5 m, trying to work with a microscope proved tricky indeed! 'Sometimes we'd collect so much stuff – buckets and buckets of small jelly-like salps, or off the north-west, brown algae that grows in slicks. We'd collect that too and have these buckets of brown goo we'd have to sort through.'

CSIRO is one of only a handful of institutions internationally that have undertaken this type of data gathering, and at the time it was a big deal. 'It was quite a high-

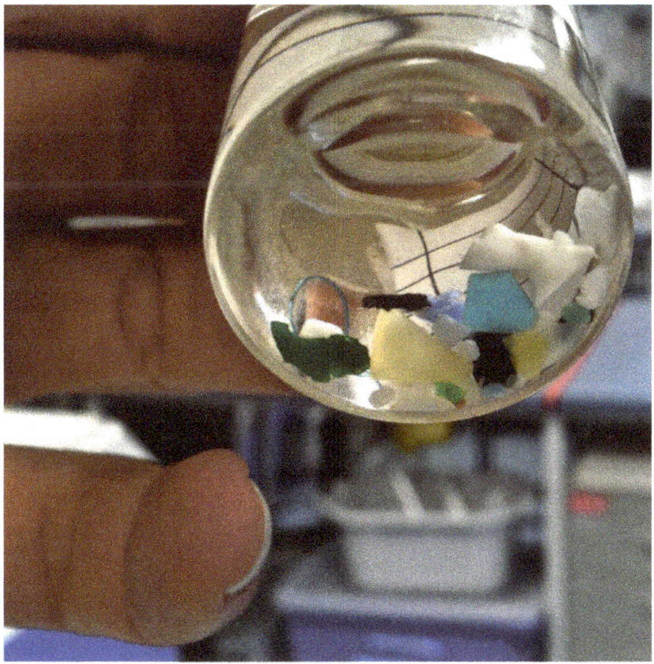

A sample of plastics from the oceans around Australia. Source: MNF/Julia Reisser.

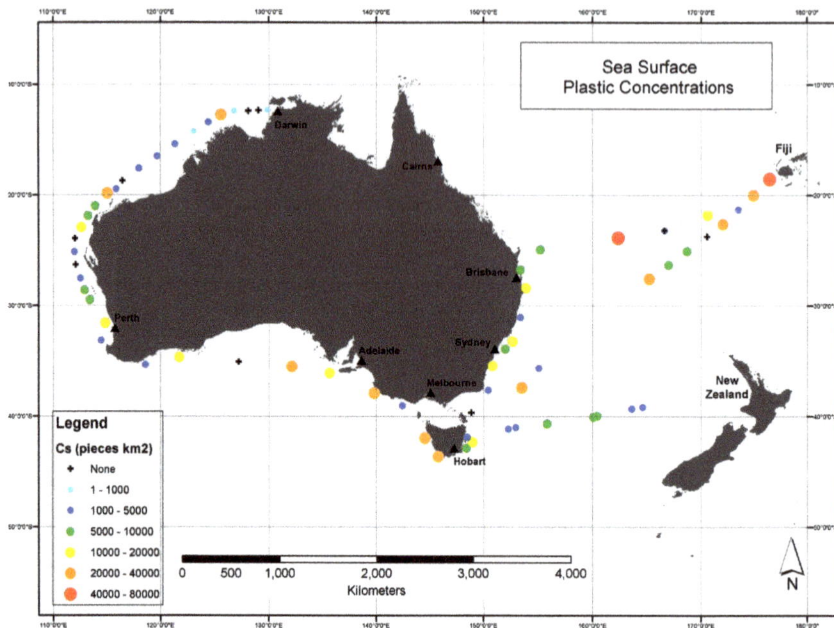

The distribution of floating marine plastics found in the oceans around Australia. This research was conducted by PhD student Julia Reisser, who was allocated sea time on several transit voyages. Once primary voyages are allocated, transit voyages are offered to early career researchers and students to undertake research. Source: Julia Reisser.

profile undertaking', says Chris. 'Our project leader probably did a couple of hundred interviews.' The team and its findings were reported in all major Australian newspapers, and even made it into the *Wall Street Journal* and *National Geographic*.

But when they published, not everyone was happy. 'We got a lot of pushback from people', says Chris. 'We backed it up with modelling but they were quite surprised by what we'd found.' The controversial conclusion Chris and his team had come to was, in essence, that most of the plastic found in Australian waters comes from, well, Australia. Their methodology sounds quite simple. In their circumnavigation of the continent, they found that the closer they came to the cities, the more plastic they encountered. 'It's striking how strong the pattern was', says Chris. 'It's hard to imagine any other mechanism that would drive that. There's this image out there that Australia's pretty clean and that all this stuff is actually coming from other places – from Indonesia, other parts of Asia, blowing across the Indian Ocean from Africa etc., but no.'

It's hard to imagine more important, indeed urgent, work than that being undertaken by Chris and his team from CSIRO and the University of Western Australia onboard *Southern Surveyor*. But it's not simply a gathering of data that drives them. Chris would dearly love to help realise some of those changes of attitudes, indeed of human behaviour, which may lead to a reduction of the waste we so carelessly throw into the sea. 'We're essentially trying to do something all the way end to end, look at the sources, look at the impact on wildlife, then look at policy solutions.'

Chris anticipates the arrival of *Investigator* but will always appreciate the time with *Surveyor*, old though she was. 'I am not one of those who doesn't get seasick', he says, 'and she was pretty kind, even in rough seas. But you know', he adds with a note of curiosity, 'there's a stinky spot onboard that ship.' This intrigues me. 'As you come down from the upper deck, just at the bottom of the stairs as you hit the linoleum. There's this wretched smell. I used to have to hold my breath every time I came down those stairs.' What kind of smell? I almost hesitate to ask. Chris thinks for a bit, before a slightly queasy expression comes over him. 'A mix of … vomit and rotting garbage, a slightly human, acrid smell. Perhaps I don't want to know', he says. I agree that's probably wise.

Brian Griffiths

Marine biologist

> *'I can remember one night being rudely awoken by a crewman saying, "Don't you realise there's a fire alarm going on?" I'd slept through it!'*

Having had history with *Southern Surveyor* right from the early days after her purchase by CSIRO's Fisheries Division, marine biologist Brian Griffiths well remembers his reaction upon seeing her for the first time, not long after her arrival in Australia. 'I was astounded at how much bigger she was than *Franklin*', he says. 'She was in remarkably good nick when she first got here, but not that suited to the work we wanted her to do.' As part of the Fisheries Division, Brian was involved in *Surveyor*'s initial conversion in Launceston where 'various bits of her were ripped out, and various other bits put in'. It was here her onboard chemistry and other laboratories as well as the CTD room were built into the superstructure, enabling the next incarnation in the life of this remarkable vessel to commence. However, the vagaries of science being what they are, says Brian, by the time the conversion was complete the direction of CSIRO's science program had shifted somewhat. 'That's the difficult thing with research ships', he says, 'trying to keep up with the pace of change. It's much easier in a land-based laboratory, but with a research vessel, it ain't so simple!'

Coming from a small Canadian town halfway between Prince Rupert and Vancouver on the British Columbia coast, it's not surprising that Brian ended up spending a good deal of his professional life onboard ships. 'Just how small was this town?' I ask. 'Well, we were about 300 miles from Vancouver and the only way you could get into it was by boat or aeroplane', he explains. 'We had a lake up behind the town and the ocean in front of us so we basically grew up in boats.' As a kid, there wasn't much science in his life to speak of, either. His was a pulp and paper town, basically owned by the company, where his father was the paymaster at the local mill.

As a young undergraduate from the University of British Columbia, Brian hatched plans with a friend to come to Australia for some diving on the Great Barrier Reef. A couple of weeks before making the trip, however, his friend decided to cancel and get married. 'I decided to come anyway', says Brian. Landing in Sydney, broke, he scoured the papers and spotted a job in medical research. 'I rang up on the Monday, was interviewed on the Tuesday, started on the Wednesday and met my wife on the Thursday!' Not a bad welcome to Australia, I suggest.

After two and a half years with a new job and a new wife, in 1970 Brian was offered a position in CSIRO's Fisheries and Oceanography Divisions as the curator of zooplankton collections, which was even then extensive. 'There was a lot to curate', he says. One set of samples which his boss, David Tranter, had collected swimming in a lake in Israel was preserved in whisky! 'He assured me it was the only alcohol he owned!'

Those first few years with CSIRO Brian devoted to expanding and extending the zooplankton collection even further. For this purpose he and David Tranter developed several sampling systems, based around specially designed nets, to look at plankton at different depths. 'This was how you got a picture of the way plankton is distributed in the water column', he tells me. 'You take a sample at the surface, then another at say, 50 m, then another a hundred etc.' Two nets were developed, one for the shallower inshore waters off Western Australia and a larger net for the greater depths of the east coast.

Then, with a change of emphasis in research programming within CSIRO (something that seems to happen frequently throughout Brian's career), he found himself becoming an estuarine scientist, working on *Southern Surveyor*'s predecessors, *Soela* and *Franklin*. 'We then began to look at how some of the intermittent estuaries along the east coast of Australia worked', he says. 'Sometimes they're relatively fresh due to rainfall, sometimes they're quite marine.'

Brian spent several years examining the community structure and distribution of tiny marine creatures in estuaries, particularly in the region of the drowned river valley of Port Hacking on the borders of Sydney's Royal National Park, a very interesting place. 'There are various bays running off Port Hacking – 20–30 m deep – which are regularly flushed with clean water', he tells me.

The ten or so voyages Brian undertook with *Southern Surveyor*, beginning in the early 1990s, he remembers as 'immensely tiring but ultimately successful'. Always he was struck by how comfortable she was in rough weather. 'She was a dream in the water', he says, fondly remembering her 'lovely slow, very predictable roll when she was broadside on'. This predictability made her very easy to work in. *Franklin*, by contrast, was completely unpredictable and not nearly as pleasant a ship to be on, he says.

'We did a lot of very good work on *Surveyor*', he says, the bulk of which was carried out around Tasmania, looking at the diets of marine organisms on the edge of the continental shelf. By then a biologist with considerable experience with zooplankton, Brian tells me that due to yet another change of emphasis within the organisation ('change is just what happens in science') he was asked to 'move down a trophic level' to begin looking not at zooplankton but at smaller phytoplankton. It was an opportunity he would make the most of.

'We had very little information about the seasonality of phytoplankton production', he tells me. At this point, I suspect I had one of those looks on my face that told him I was struggling to register the notion of phytoplankton and its undoubted relevance in the wider scheme of things. 'It's really the basis of all the aquatic food chains', he says, 'and also part of the basis of carbon dioxide cycling from the atmosphere into the ocean.'

Brian explains that phytoplankton 'fixes' carbon from the ocean, the zooplankton eats the phytoplankton, then other things eat the zooplankton and so on up the food chain. 'We really didn't know very much about the cycles of production of phytoplankton around Australia', he tells me, before – like any good teacher – coming to my rescue with an analogy. 'Think of your lawn as phytoplankton. In winter when it's cold and dark it doesn't grow much, but in spring it goes like crazy, especially if you fertilise it.' Some work was apparently done in the 1950s and 1960s, he says, but it was pretty sparse.

'I developed a series of methods for estimating how much carbon the phytoplankton were taking up out of the ocean', he says. Other questions were also addressed. What was eating phytoplankton? Where does it go? Does it sink to the bottom? Does it get recycled? Just how the work was carried out is a fascinating story in itself.

Brian's sampling methods involved oceanographic sampling along transects running perpendicular to the continental shelf, sailing out into the deeper water. The team used CTDs and Niskin bottles which would take water at various depths from the upper end of the water column. Radioactive carbon was added to the samples, which were then incubated for an hour under different light intensities, designed to replicate various depths of the ocean. Any unfixed carbon was removed, enabling the amount which had been taken up by the phytoplankton to be measured. 'Replicating the ocean's light regime like this', he says, 'enabled us to get a "carbon taken up versus light" curve, which in turn allowed us to model phytoplankton at any light intensity in the ocean.' It also offered an understanding of phytoplankton production at various parts of the water column. 'By doing this at a number of depths, we could understand how much phytoplankton production was going on in the upper 100–150 m of the water column.'

It was an involved business. 'I had an unusual method', Brian says. 'I would put 7 mL of seawater into a glass bottle, with twenty-one bottles for each depth sampled.' Artificial light intensities, from 'very low to quite bright', would illuminate this large and complex array of glass, recreating the position of the sun at various times of the day. Brian found that at low light intensities the plankton would take every bit of light they

A photo taken from the ship's workboat out in the Pacific Ocean on a voyage led by Richard Arculus from the Australian National University. Source: MNF/Richard Arculus.

could, 'but in bright light you reach a point where they just can't handle it'. Seven millilitres, I point out, is not very much water to work with, but he tells me that his technique was sufficiently sensitive for him to get 'really great numbers'.

'I'm confident', he believes, 'that we were measuring real rates, and I think we have a better understanding of how the fisheries work on these slopes.'

Brian's *Surveyor* voyages 'went mainly very right. We had experienced sea-going personnel onboard who knew their jobs and were fun to be with'. But working up to seventeen hours a day took its toll, particularly in rough weather when sleep deprivation became the norm. So when he crashed, he really crashed. 'I can remember', he says, 'one night being rudely awoken by a crewman saying, "Don't you realise there's a fire alarm going on?" I'd slept through it!'

He can also remember nets torn away by becoming too 'full of sponges and things', waves crashing through open doors of the CTD room and cascading to the decks below, and losing trawling equipment after the crew had determined that it was stuck so fast to something on the bottom that all there remained to do was simply cut the cable. 'A heartbreaking moment', says Brian, 'but it's all part of being at sea. My attitude to things we put over the side – if we get it back – is "Great." When things go wrong you just have to shrug your shoulders and say, "Okay, how do we recover from this?"'

Like so much of the work of marine science, rarely was a 'Eureka' moment to be had at sea. Brian's results could only be gleaned long after the voyage's conclusion and the myriad amounts of data digested and analysed, with findings eventually published and added to the general lexicon of scientific knowledge. That's not to say, however, that there weren't what he calls 'some really neat moments when you see things that are extraordinary'. At night off the east coast of Tasmania, for example, he saw large luminous patches of 'wonderful soft blue phosphorescent light', emanating from organisms called pyrosoma. 'Acres in extent', he says. 'Fantastic to look at.'

Rudy Kloser

Marine biologist

> *'Counting fish is a bit like counting trees, except you can't see them and they move.'*

Rudy Kloser gives me a completely new term to add to my vocabulary – 'marine acoustics'. Not a new form of music harnessing the untapped tonal intelligence of sea urchins but, put simply, a great way to count fish. 'We send out sound waves which reflect off structures in the water column and usually they're fish', he tells me. These sounds can be 'tuned' to bounce off certain parts of the fish mass, and from the 'backscatter' sound, the biomass can be determined. Rudy has developed a series of techniques for this process, helping to gather important information for fisheries management and sustainability. Just how accurate, I ask, does he consider it to be? 'Right', he says slowly. 'That's a very good question. The trouble is that counting fish is a bit like counting trees, except you can't see them and they move.'

'The biomass of deepwater fish stocks' is how Rudy sums up his enduring interest in the oceans, and over the course of his many voyages with the *Surveyor* – roughly two a year over fifteen years – he would become well versed in determining the status of many species. One in particular, however, the controversial orange roughy, would define his career. And it starts right at the beginning of the ship's life with Fisheries.

'I was involved right at the start making sure *Surveyor* had the right acoustics onboard', says Rudy. In the Launceston refit of 1990, he oversaw the installation of the state-of-the-art, Norwegian-built Simrad EK-500 Echosounder or, more simply, fish-finding sonar – one of the earliest ever built, he tells me. He took it out on the Tamar River on *Surveyor*'s maiden voyage down the east coast of Tasmania to Hobart. 'It sends out a pulse of low-frequency sound', says Rudy. '38 kilohertz suited to long-range activities – a narrow directed beam which simply reflects back. Everything's calibrated so you know exactly how much sound you put out, and how much you receive.'

This voyage was the same one that Matt Sherlock had described to me, as the *Surveyor* weaved its way at night through myriad fishing boats lined up to have their

Attached to the hull of *Southern Surveyor* was a gondola that contained some of the ship's transducers and receivers for the acoustic systems. This included three echo sounders that together could map fish populations and distinguish one species from another as the ship steamed along. Source: CSIRO.

'shot on the hill' in these rich fishing grounds. Rudy was also busy, directing the ship to make a survey to give the new hardware a run. 'It was during the heyday of the orange roughy fishing era', he says. 'I remember seeing twenty or thirty boats all fishing over a small area, and here we were, this big large boat doing transects among them, counting the fish.' If the men on the boats eyed the big blue hull of *Surveyor* with foreboding as it slowly weaved its way among them, they had good reason to.

Besides the echosounder, *Surveyor* was sporting another piece of futuristic equipment, the then revolutionary Global Positioning System. Today we take this remarkable technology for granted, using it daily on our smartphones to find a chemist, a pub or a quicker way home. But in 1990 it was a very big deal, despite its limitations. 'GPS had only just come out and it was a big thing, costing thousands', says Rudy. 'But there was only a short window of a few hours a day when you could actually use it, and the equipment was very cumbersome.' Far different from the hand-held convenience of today, 'a great big box' gave a series of numbers which then had to be put through a huge hard drive. 'It was all very complicated', he says. 'Looking back it's quite amusing now.' There was, however, very little that was amusing when Rudy put some of the new equipment to work, embroiling himself in the controversy of the orange roughy.

A large deepwater fish found around seamounts and canyons between 700–1000 m, the orange roughy had remained relatively unmolested until the advent of readily accessible echosounding equipment in the late 1980s, at which point, says Rudy, a fishing frenzy began. 'The aggregate was only discovered in 1989, so it was a brand new fisheries industry and it was like an explosion.' In just two years, he says, the un-fished orange roughy stocks off the east coast and south of Tasmania were reduced by more than half. 'It was virtually open slather.'

The fish was particularly vulnerable to overfishing because of its slow rate of growth and exceptionally long lifespan of up to 150 years. 'The industry were getting better equipment than the scientists', says Rudy. 'And the more money that could be made with the fishing, the more gear could be bought to further harvest the resource.'

In order to gain a clear picture of this new resource, it was decided to make a proper estimation of its size. In *Surveyor*'s operations room, Rudy began transects back and forth across one of the orange roughy's primary breeding grounds, St Helen's Hill, which, as Matt had explained, was a large underwater seamount rising from 1100–600 m below the surface. Watching his screen in the ship's operations room, taking into account interference and carefully ensuring the quality of the data he was looking at was high, Rudy made a two-dimensional map of what was beneath. 'You'd see the hill rising and falling, and beside the hill you'd see these echoes, which were the schools of orange roughy', he says. Shadows and numbers on a screen, however, did not lead to an instant result. Many more weeks of work were required back on shore for conclusions to be drawn. 'Computing wasn't quite what it is today', he tells me. 'There was a lot of data we were collecting for the time and it was months spent crunching the numbers.'

But when those numbers did start to come in, Rudy knew it was information that many people were not going to want to hear. His estimates of orange roughy stock size based on the biomass began to form at much lower numbers than what was popularly believed. 'There had been rumours going around that there could be up to a million tons of orange roughy out there', he says, 'but I was only coming up with 100 000. That, I knew, was controversial.'

The accumulated data and evidence was outlined slowly and methodically in various internal meetings – other elements such as the age of the fish were factored and weighed ('There was a lot of double checking of figures, I can tell you!') and the science was allowed to speak for itself. All of it pointed towards a massive contraction in the resource having occurred over a very short time.

'We made recommendations to put in some fairly large reductions in quotas', he says. Media attention was understandably high, and although Rudy was spared the blowtorch of television he was nonetheless front and centre at some rather heated public meetings attended by scientists, fishermen and fisheries management. 'Yes, I got called lots of things!' he says, 'It got controversial, but we held the science course.' It

Fisheries research often involves collecting small samples of fish from a region. This is a catch of fish from the Bass Canyon, which was collected during a transit voyage. Students were given the opportunity to use different sampling equipment to collect deep sea fauna. Source: MNF.

was, however, no bloodbath and slowly, despite what was for some an unpalatable message, the science came to be accepted and recommendations were implemented. 'I look back at that time fondly', he says. 'You were relevant, you were affecting someone's life and you wanted to make sure you were right. A review from someone whose life you are affecting is always the best review.'

Rudy believes this work resulted in not only an eventual improvement in the fish stock, but of the industry as a whole. 'Management started to take hold of the industry and our results started to be used in practice.' The tools that were developed to assess the stock have also become part of the industry. 'Now we do a lot of our work on fishing vessels, calibrating the equipment and they go and do the surveys themselves', says Rudy. In 2006, the orange roughy was the first commercial fish stock to be added to the Conservation Dependent list, and it remains so today.

Now, says Rudy, the stocks of orange roughy (which I learn is actually more brick-red than orange) are on the increase and to him that is the most satisfying part of his work. I ask Rudy if he considers himself the person who saved the orange roughy species. He laughs, a little embarrassed, but admits, 'Look, I have actually wondered about that but I think what would have happened is that they just would have become uneconomical. I think we helped to provide a mechanism where you could make it sustainable. It's a shame', he continues, 'with the value of hindsight, you could have had a sustainable fishery for generations. A lot of people made a lot of money at the beginning but the price of the fish went down so quickly that many made nothing.'

Rudy's other work with *Surveyor* included mapping of seamounts, exploring the habits of the blue grenadier, and a series of voyages off the Tasmanian and Victorian coastlines concerned with the overall marine ecosystem. This entailed developing seafloor and habitat maps which, he says, 'helped form the backbone of information that went towards the setting up of marine protected areas'.

A successful voyage was not something that simply happened, says Rudy. It needed to be worked on, and the crew were often the key. 'We were trying a lot of new gear – new winches, deep towed bodies, new handling systems, and we needed a crew that knew how to handle all of it.' He is also grateful for *Surveyor*'s former legacy as a fishing vessel, as his own work required similar skill sets.

The *Surveyor*'s former fishing skippers set the tone for a happy working relationship. 'They often came from a fishing background', he says, 'and just loved anything to do with fish and fishing.' Their diplomatic skills were also needed at times, especially during the orange roughy controversy when 'there wasn't much love out there. All through the '90s we'd be out there with the fishermen on the same grounds, and you'd have to talk to them and let them know what you were doing. Our masters were terrific in knowing just how to talk to them, because, for a while back then, it was a bit like the wild west.'

MNF SHIP'S GROUP

Lisa Woodward

Voyage manager

'I just wish I'd found this job twenty years ago.'

Lisa, you might say, came up through the ranks. Seven years ago, having spent her career working with young postgraduate doctors, she decided a change was in order and took up a one-year admin position as Assistant to the Director on the Marine National Facility. So capable did she prove herself that the powers-that-be found the money to make her permanent. Then one day her boss ('Bless his heart', she says) took her aside. 'I think you'd do well at voyage operations', he said, to her complete surprise. 'So, with no ocean-going experience whatsoever, but with enthusiasm and an eye for detail, I went to sea.'

From voyage operations, Lisa continued her rise and in a short time found herself a MNF voyage manager. 'You must have had good sea legs', I suggest to this cheerful and down-to-earth woman. 'So it proved', she says, 'but I didn't even know that when I went to sea. I hadn't even been on a yacht!'

Lisa's first experience on *Southern Surveyor* was a voyage from Mackay to Newcastle, and the weather promised to be atrocious. Being a person who gets queasy looking at a model boat in a glass case, I wondered how she fared on that maiden voyage. 'I just loved it', she says brightly. 'The sea was pitching but I was up the front saying, "Go faster! Go faster!"' With Lisa and the sea, it was love at first sight and she could only reflect on what she'd been missing all her life. This is despite what sounds like a somewhat unpleasant 'initiation'.

'In my cabin one day', she says, 'I found my bed had been beautifully made. There were chocolates on my pillow and next to it a bottle of (non-alcoholic) champagne.' The cook at the time was bald, loud and rather intimidating. 'If you saw a pan flying out of the kitchen, you stayed away because it meant he was in a bad mood.' While contemplating the treats arranged on her bed, the second cook appeared at her cabin door and nonchalantly mentioned a barbecue on the back deck in honour of her first voyage. 'Lovely,' said Lisa, 'I'll be down in a while.' 'Actually', he said in a

Southern Surveyor tied up at Cairns port in preparation for a voyage through the Great Barrier Reef. Source: MNF/Matt Chamberlain.

slightly more emphatic tone, 'you should probably come now.' Lisa began to smell a rat but, as she says, 'What could I do?'

There on the deck was the cook, looking just as intimidating but dressed bizarrely as the god of the deep, King Neptune, and arranged around him the entire ship's crew. After a short ceremony in which some curious text was read from a scroll, a bucket of 'something horrible' was tipped over Lisa's head. She had been initiated in the true maritime tradition for all those crossing the equator for the first time. Now she was, well and truly, one of the boys. 'Actually, I've always been one of the boys', she says, laughing.

The voyage manager is, officially at least, the MNF's representative onboard the vessel. Unofficially, they're the eyes and ears of the ship and the company as well as the person who has to monitor relationships, moods and stress levels, listen to everybody's complaints, smooth over arguments and generally make sure things get done. 'To assist in making the voyage as pleasant, productive and as safe as possible', is the diplomatic spin Lisa puts on the job description. I suggest the term 'fireman', at which she smiles, but the words that perhaps best describe the essential quality required is 'people skills'. On a working sea voyage these, I learn, are well and truly put to the test.

Lisa completed no fewer than eighteen voyages on *Southern Surveyor*, the shortest a couple of weeks' duration and the longest lasting a month.

A ship, Lisa tells me, tends to act as an amplifier of human behaviour. 'Even small things such as taking someone else's washing out of the machine and not bothering to put it in the dryer. After a few days of things like that – with some bad weather thrown in, well, people can get extremely upset in ways they wouldn't at home.' Part of her job was to keep these things in check. 'At sea we even tend to speak a different language', she tells me. Emails, particularly work emails, become scrutinised for meaning and inferences which simply aren't there. People who are clashing feel hemmed in, airing their grievances to others rather than confronting the person with whom they have the issue. Hence, miscommunications and misunderstandings, says Lisa, can take off in a cloistered environment such as on the *Southern Surveyor.* One of the rules of the job is to 'deal with the stuff that happens at sea, *at sea*' rather than take it back to shore.

Like almost everyone on the *Surveyor,* Lisa worked a twelve-hour shift but in fact was never entirely off-duty. With only one voyage manager, if something happened, day or night, Lisa's onboard phone would ring. Luckily, she has never been affected by seasickness and, unlike many, sleeps soundly onboard. 'Some people just couldn't get into the rhythm of sleeping on the ship and every little noise would wake them', she says. 'I didn't realise how much seasickness and lack of sleep can impact on people's moods and perceptions.'

A typical day for Lisa would begin with an early-morning trip to the ship's small and somewhat rudimentary gym. Then, like a doctor doing the rounds in a hospital, she would bring herself up to speed with what had happened overnight. First the operations room, where she would assess a tired group of scientists and students 'and decide whether they needed a prod or a cup of tea', then a check on how the night operations had fared, whether equipment had failed and so on. Then it would be up to the bridge to consult the Officer of the Watch before his 8am relief, followed by breakfast in the galley. Here, she says, a lot can be gauged by the conversations of scientists and crew. If there had been any issues overnight, Lisa dealt with them before meeting with the Chief Scientist and Master to discuss the day's schedule. 'And if the weather was bad', she says, 'we didn't do anything except hunker down.'

Lisa's enthusiasm for the job she performed onboard *Southern Surveyor* is infectious. I ask her to name the highs and her lows of her many voyages. The latter are rare, though she claims to have had more than her fair share of medivacs and witnessed the infamous loss of the $250 000 SeaSoar. In fact, in a peculiar way, Lisa feels somewhat complicit in its demise. Having one day decided to relieve one of the students in the control room monitoring the SeaSoar's receiving screen, Lisa watched the data forming patterns when, like a patient in a hospital on life support, it all suddenly flatlined. 'I said to the scientist, "Er, that's not good when it goes like that is it?"' She couldn't have been more correct. A short time later, the broken towing cable with a couple of severed wires poking out its end was winched, dismally, back onboard.

The sun sets over the Pacific Ocean. Source: MNF/Richard Arculus.

Despite these rare instances, Lisa's life, she says, has been greatly enriched by her time at sea. And why not? Perks of the job include sailing past live volcanoes in Vanuatu where the glow of lava lights up the night sky, or being taught seafaring from an old hand such as First Mate, Boysey. 'I have no maritime background at all', she says, 'so he took me aside and taught me things like navigation and coastal mapping. He'd even give me homework. Once he got me to chart a course right into Hobart harbour', she adds, still a little incredulous. 'He was just a wonderful teacher.'

Then there are the countless close encounters with seabirds, dolphins and, though it took a while, whales. 'I didn't believe they actually existed, it took me so long to see one!' Apart from these memories, Lisa tells me she simply loves being aboard the ship. 'Not everyone liked sailing on the *Surveyor* but I just love it', she says. 'She was a wonderfully stable sea ship.'

But there was something else about her for which Lisa will always be thankful. 'You always felt very connected to the sea on the *Surveyor*', she says. 'Wherever you went on her, you always had a good view of it and a good sense of it.' She's not as sure the new *Investigator* will be quite so homely. Her only regret? 'I just wish I'd found this job twenty years ago.'

Ron Plaschke

Director, Marine National Facility

> *'There was a saying onboard – when it comes to doing CTD stations, the first ten thousand are the worst.'*

Standing over 180 cm in his socks, Ron Plaschke, the future head of the Marine National Facility, stood slightly taller than the average marine chemistry technician when he joined CSIRO in 1984, so contemplating the often low-slung doorways into some of the ships' laboratories was a particular issue. 'I went to sea on a whole range of vessels', he says, and in his time he seems to have indeed become acquainted with just about everything CSIRO put to sea. This includes the *Franklin* and *Southern Surveyor*, as well as the Australian Antarctic Division's *Aurora Australis*. Ron was the first person I had met who had worked on the French-built stern-trawler *Soela*. 'The average height of a Frenchman being what it was when they designed *Soela* meant I had to duck down a lot on that ship', he tells me. 'In the chem lab, there was one small part of the ceiling which was raised to accommodate air conditioning. It also happened to be the only part of the lab where I could stand up straight!' Ron also recalls the ironically named *Sprightly*, 'a World War II-era ocean-going tug, semi-converted for research purposes'.

These were the days before the advent of the CTD, now one of the standard tools of oceanographic research. 'Back then', says Ron, 'we just had a wire which we had to screw the sample bottles onto. You'd be hanging off the end of what they called the "hero platform", screwing on metal Nansen bottles to a cable'. With every wave and every roll of the ship, Ron could be up to his waist in water. Slightly perturbed, I enquire whether he was tethered to anything. 'Er, no', he says. 'Health and safety has evolved quite a lot since then. They didn't call it the hero platform for nothing!'

The Nansen bottle system sounds quite ingenious, requiring a small metal weight called a 'messenger' to be released at the top of the wire, which itself could be plunging thousands of metres to the seafloor. When the first messenger struck the first primed bottle, it would cause it to flip over, capturing a sample of water, which would then release another messenger. The process would be repeated with all the many bottles strung along the length of the cable. 'I've still got some of those old messenger weights in my office actually', says Ron. 'Looking back at it now it was all so laborious.'

Citing the advent of satellite imaging as one of the great leaps forward in marine science research, Ron remembers what it was like beforehand, particularly when trying to locate newly discovered eddies. 'Australian oceanographers were one of the first to discover eddy systems', he tells me. 'We used to run around looking at them but it was a bit like flying blind. We knew they were there but we didn't have the big picture. When the satellite imagery came in people said, "Wow look, we're right on the edge of one!" It opened up the whole picture.'

Ron, like many, describes *Southern Surveyor* as a 'great old boat with a capable, well-proven hull', which would be continually tested on the unforgiving seas south of Australia. On one dramatic trip he remembers encountering a 'particularly unusual swell', uncharacteristically 'short' (i.e. with the waves close together) for that part of the ocean, which seemed for periods to be in sequence with the roll of the ship. Carrying out CTD operations with the wet laboratory open to the sea, Ron was in his cabin trying to chase down some of the sleep which had thus far largely eluded him. 'We were making these synchronous rolls which were keeping me awake', he says. 'Every time a wave came it seemed up push us up a little bit higher out of the water and we'd roll some more.'

At one point a combination of swell and roll, and then another, this time deep and sickening, nearly threw Ron out of his bed. 'Wait …', he says to me, pausing, 'maybe it actually did throw me out of my bed.' He's in no doubt about the sudden blaring of the fire alarm and a sense that 'something terrible had happened'; he leapt up, struggling to put on his overalls and life jacket in his sleep-deprived haze.

'As I clambered up to the deck', he recalls, 'water was pouring down the stairs in front of me, more water was sloshing down the alleyway of the main deck and, as I glanced into the wet lab, all I could see was chaos.' The awful synchronicity of swell and roll had timed perfectly to force the *Surveyor* to dip side-on into the maw of a particularly fearsome wave. 'It completely flooded the wet lab as well as the people who were in it', he says. It's a story I have heard previously, but from another perspective. I had heard of Cathy Bulman's close encounter: how, washed off her feet by the incoming rush of water, she was grabbed just in time by a crew member. Ron tells me that she was also in danger from the heavy CTD coming back down onto the deck, which could easily have hit her. 'So it was a bit of a close call', he says. 'Of course, that was how things happened back then. It wouldn't happen that way now.'

On his fifteen or so trips on *Southern Surveyor*, Ron was kept busy as a hydrochemistry technician keeping the laboratory running around the clock, setting up the CTD and its sample bottles, analysing their water once they had come onboard, and training others to do so as well. Then he spent hours locked in the chem lab

Southern Surveyor facing a 3 m swell with winds of up to 30 knots in waters 100 km west of Darwin. Source: MNF.

undertaking salinity, oxygen and nutrient analysis. I'd wondered what this analysis entailed, and Ron provides some of the finer details on how ocean water was examined onboard *Southern Surveyor*. 'First we'd sample for salinity measured in a small bottle that looked like a small brown stubby', he explains. 'That would have to be rinsed and filled three times. You'd then sample for dissolved oxygen, which involved adding chemicals to "fix" the oxygen, then titrate the sample to determine the amount. Little plastic tubes were used to collect nutrients, such as phosphate, nitrate, silicate and so on', he says.

In the early days, Ron tells me, this data was all taken down by hand which, particularly in bad weather, was both time-consuming and exhausting. 'You'd be in the chem lab mixing up chemicals and reagents and sometimes the weather doesn't let up', he says. 'On the other hand, there's nothing else to do out there anyway, so the notion of twelve- to fourteen-hour days didn't seem as bad as it would otherwise.'

The great difference between *Southern Surveyor* and some of her predecessors, Ron tells me, was the variety of the work. He's lost count of the number of transect trips he's done across the Tasman Sea on *Franklin*. 'There was a saying onboard', he says, 'when it comes to doing CTD stations, the first ten thousand are the worst.' He chuckles grimly. Indeed the tedium of the routine sounds daunting. 'You stop, you do a CTD station which takes an hour per 1000 m of water, so you're on station for a number of hours. Then you bring it up, get going again, analyse the samples while you're steaming towards the next location, at which point you stop again, put it back in the water and do the whole thing all over again. And that can go on, non-stop, seven days a weeks for three or four weeks, every thirty nautical miles. That's what really tests people in terms of resilience', he says.

But at least on *Southern Surveyor* there was more variety. 'We'd do some CTDs, but then we'd go off and do some fishing', he says. 'So you'd get to talk to some of the marine biologists and sort fish from the deep ocean, some of which were amazing creatures you'd never get a chance to see anywhere else.'

The evolution of the relationship between the crew and scientists has been one of the more noticeable measures of progress over the course of Ron's career. 'When I started to go to sea in the mid 1980s', he says, 'a lot of the marine crew were simply not used to working with other people on the vessel. Some of them viewed you as the enemy who were coming to take over their ship, which was hard to deal with at times.' It's a different story today, although perhaps the change of attitudes has merely reflected those of society as a whole. 'Today, there's no drugs or alcohol onboard and the ship operates as one big team.'

The level of duty of care has also evolved, including towards those with seasickness. 'When I started going to sea and you were seasick and couldn't eat, well, bad luck', Ron says. 'They would only consider putting you off if the situation became desperate. Today, if someone has been seasick for more than two days, people are worried about them, emails to home are fired off, and conversations are begun about where they could be put off the ship. It's an interesting issue for us to manage.'

In a wider sense, Ron points out how the nature of the Marine National Facility too has evolved. 'We've really gone from the use of smaller, virtually single-purpose vessels to something much more multi-disciplinary', he says. 'Today we encompass biologists, oceanographers, geoscientists, atmospheric scientists, people looking at ecosystems, whale-watchers and everything in between.' As for Ron's own evolution – his rise from hydrochemist, toiling away inside the ocean-going laboratories of the various CSIRO and MNF vessels, to eventually running the organisation himself – that's a story in itself.

In 1984, when a youthful Ron started going to sea, it was an adventure. Then, as his young family grew, he began to indulge in that singularly dangerous practice which any sailor spending large amounts of time away from home and family should forbid themselves under any circumstances – thinking.

'Sometimes I'd be doing back-to-back trips of three or four weeks at a time', he tells me. 'And suddenly, it got quite tough going to sea.' His wife, I suggest, probably didn't find it much fun either. 'Yes, that's what I mean', he adds. 'It was tough on everybody. There's a lot of time out there spent doing routine work, and your brain has a lot of space to think.' Somewhere deep in the Southern Ocean, Ron had his moment of revelation on the road to Damascus. 'One day I just asked myself, "What's the

Southern Surveyor tied alongside the CSIRO Marine Laboratories in Hobart. In 2014 she was sold following an international tender process. Source: CSIRO.

point of coming out here and being miserable? Either get out, or find a way to enjoy it." It was quite an epiphany', he says.

Although committed to his beloved realm of marine science, Ron, when not tied to the laboratory, started to take an interest in the wider picture of the MNF, leading eventually to a position as Operations Officer in the ship's management group. 'I'd then go to sea more as a voyage manager than a hydrochemist, and I went to sea less often', he says. Ron's new course was set, and five years ago he ascended to the chair of Director of the Marine National Facility.

What he remembers most about his time at sea are the friendships. 'Occasionally you get a spark of personalities that can cause an issue', he tells me, 'but the vast majority of people develop life-long friendships at sea. You need to work together, and you're all in the same boat, literally.'

It's also vitally important, he continues, that people enjoy their time at sea. 'I have a colleague', says Ron, 'a master mariner who makes a pact with himself every time he goes up the gangway that he's going to have a good time. Because if you don't you can end up very miserable. And most people do have a good time. At the end of a voyage, you come into port and have a meal, have a couple of drinks – finally – and tell some stories together. It's one of the things you look really forward to.'

As Ron and I talk in the sun-filled cafeteria of the CSIRO's Marine Laboratories in Hobart, we occasionally glance across the water to the gleaming blue, green and white hull of the brand new RV *Investigator*, towering majestically above Macquarie

Wharf, preparing for its trials. I mention that it looks a far cry from the *Soela*. 'Oh, yes, the *Soela*', he exclaims. 'That was something else I remember about her. You could tell what part of the ship you were in even with your eyes closed – by the smell!' We begin laughing but he assures me it's not exaggeration. 'If you started aft, there were the toilets, then you'd come slightly forward to the formalin smell of the fish lab, then the chem lab had its own aromas, then a waft from the engine room which mixed with the stale cigarette smells from the smokers' room, which in turn mixed with the food smells from the galley.' The seasickness bags on this vessel, I assume, were always well utilised.

I'm assured that smells won't be an issue on *Investigator*, nor does it seem likely that Ron will have trouble passing through her doorways.

Don McKenzie

Operations manager

> *'When I recruit someone for my team, I choose them on how good they are with people, rather than how much they know about marine science. You can always ask someone about the science.'*

Needless to say, every one of *Southern Surveyor*'s dozens of voyages had to be meticulously planned, and every one of them was different. 'Even when you were sailing with the same people you'd been with on earlier voyages', says Don, 'it could all be completely different.' Don was one of the people whose job was to find out what the scientists who had been given use of the Marine National Facility intended to do with it. Then he had to help them make it happen.

The process was straightforward enough: after the MNF Steering Committee approved an application, there was a lead time of at least a year before the journey could begin. This was when Don would start to get busy. 'A voyage onboard the *Surveyor* is the equivalent of a $1–2 million government grant and can be a once-in-a-career opportunity for a scientist or a science team', he says. 'They're under a fair bit of pressure to achieve their objectives in the allocated time. We need to understand what it is they want to do and help them do it.'

A thirty-seven-year veteran with CSIRO, Don speaks quietly but with a steely resolve – just the sort of bloke you'd want to be standing next to in a crisis, or even when something simply needs to get done. 'When I recruit someone for my team', he says, 'I choose them on how good they are with people, rather than how much they know about marine science. You can always ask someone about the science.'

Several months out from a voyage, Don and his team will meet with the scientists, visiting their workplace to see how they operate, and begin the all-important voyage plan. What will be required? Will it be labour-intensive? What sort of equipment do they intend to bring onboard? How will it interface with the *Surveyor*? 'Running cables, getting different amperage power to the back deck, connecting data cables from the foremast to the ops room, being able to protect a Perspex aquarium from the waves on the deck' is a tiny selection of what might be required when setting up for a new tenant on *Southern Surveyor*. And it doesn't stop when it gets to sea. 'Then you have to work out everything from getting rid of radioactive waste in a far-off port to finding dry ice in the Pilbara', he says. 'You get to know the security people and the freight people and the fuel people in all sorts of places.'

Don was also one of *Southern Surveyor*'s voyage leaders, and so was required to be completely familiar with the unusual make-up of the vessel's three-tiered personnel contingent: the almost permanent fourteen-person P&O sailing crew, the three to five science technicians and support staff, and the ten to twelve scientists. 'In a sense, we were P&O's clients, and the scientists were CSIRO's clients', he says.

In terms of sailing time, scientists were usually the most inexperienced. The crew spend 180 days a year at sea and the support technicians up to eighty-five. 'I usually liked to tell people when taking them on their initial tour of the ship, that these passageways, the mess and so on, are in fact someone's home. I think it was important to let them know that.' It's this sort of perspective that I suspect makes Don particularly fitted to the responsibilities of the job.

Before crossing into the operations and logistics side, Don began at CSIRO Marine Laboratories as a support technician for scientists. 'I worked with the new ocean colour satellites that had just come in', he says, 'looking at the amount of oxygen-producing phytoplankton, which is the start of the whole food chain.'

Working many hours collecting and processing phytoplankton and algae samples, and sorting and gutting fish, gave Don a solid grounding in the nature of working at sea. A curious contrast, no doubt, to his former life in animal production on a sheep research property near Armidale. 'I'd always wanted to be a fisherman and work on the Barrier Reef', he says. 'I never did, but at the time I thought CSIRO was the best way to get me there.'

Don's first job was onboard *Surveyor*'s predecessor, the diminutive *Franklin*, which he accompanied to the confronting latitude of 50° South. 'The ship had been shortened during the design to only 55 m long, so it had a very awkward motion and it never really behaved as it should', particularly in those 'pretty awful' southern seas. Working in an improvised laboratory fashioned out of a shipping container bolted to the *Franklin*'s rear deck, Don worked among an array of incubators, jars and bottles. 'Water was all over the place all of the time', he says. 'I was involved in filtering trace and micro-nutrients which required high levels of cleanliness. Coming from a farm background, that was a bit of a challenge, but it was a great time of my life.'

As for his time on *Surveyor*, Don recalls – as does everyone who was on the infamous transit voyage across the Great Australian Bight to Western Australia in 1992 – some very scary weather. 'I get seasick', he says, 'but usually only at the start of a trip, then it goes away.' Not this time. Don reckons he spent most of the ten days of that voyage sprawled across three beanbags in the operations room, waiting with dread for the time he would have to go back up on deck to collect samples. 'At one stage the mate changed course slightly, and the ship was briefly beam-on. She rolled so badly that everyone was thrown out of their bunks', he recalls. 'They reckoned if we hadn't come back before the next wave hit, we would have turned over. It was apparently a near-run thing.'

Things became far more pleasant up around Tonga and the Solomon Islands, however, on voyages accompanying scientific luminaries such as Richard Arculus, searching out seamounts and active volcanoes. It was on one of these voyages that Don first noticed a particular social phenomenon, based on gender. 'Science is actually a pretty gender-balanced field', he says, 'and it's far more unusual these days to put to sea with an all-male crew.' On one of Richard's voyages, however, that is exactly what transpired, and the *Surveyor* sailed devoid of a single female. 'It was the only time it happened in all my voyages', says Don, 'and from the senior scientist down, it was remarkable how quickly things degenerated into bad language and slovenly habits. Just like a boys' fishing trip, only three weeks long! We talked about it a fair bit onboard, because it was a bit of a surprise to all of us.'

'Keeping things ticking along' is how Don describes his job on the ship. Improvising if things go wrong or not to plan, making sure everyone onboard knows exactly what their role is, being the ear for everybody's myriad concerns, even satisfying the ever-changing wishes of the scientists. 'We try to identify their needs before we go to sea, but there are times when they'll just want to get, you know, a bit closer to a reef. "Do you have any charts for it?" we'll ask, but they'll always want to get *just that bit closer*'.

Charged with overseeing the daily science operations, as well as working closely with the ship's master, the deck crew and of course the scientists, Don had several insights into the social effects of time spent at sea, particularly for those unused to a long voyage. Breaking the seal of the *Surveyor*'s environment to reconnect with home could actually be counter-productive. 'On a long trip, particularly in the early days, communication was via a very expensive satellite phone. People would call up and pay $12 a minute to get loaded up with bad domestic news: kid suspended from school or a broken pump or something, and of course they were completely powerless to do anything about it.' Don says that he could sometimes spot people who had been talking with home, as for the next week or so they looked 'guilt-ridden and miserable'.

People also, he said, had a tendency to become somewhat paranoid with missives from the shore world, drawing meanings and implications that simply weren't there from quite innocuous communications – a view reinforced by Lisa Woodward. 'It's a hypersensitivity that creeps in', he says, 'partially from tiredness. *Southern Surveyor* was an old ship, and a noisy one, and with the constant movement and rolling, sleep was always hard to come by. I had a tendency to sleep well only every other night.'

Meals were a 'key part of the game', he told me. 'In the beginning when I started, you'd get seven roasts in a row, and they said the veggies were cooked longer than

Oscar Pizarro, an Autonomous Underwater Vehicle specialist from the University of Sydney, helps himself to some of the food on offer on a voyage into the Great Barrier Reef. Source: MNF/Rob Beaman.

the meat. All very different today, of course, but meals are still a vital time for people to socialise and catch up on the day's events.' On one occasion, though, the meals became a flashpoint, not for the people eating them but for those preparing them. 'We had a couple of cooks who fought terribly', explains Don, trying, I suspect, to suppress a slight note of amusement.

'One of them had to be shut away in his cabin. When we got to New Zealand, they had to be escorted – separately – off the ship and flown home! But that was very unusual.' The ship's crew handled this bizarre episode with exemplary tact, says Don, so the actual cause of the culinary stoush remains a mystery. 'All in all though, the ship is an interdependent team, and relationships have to be strong.'

With the advent of technology, times have changed. Years ago, the *Surveyor*'s rec room, with its ubiquitous beanbag chairs and racks of VHS movies, became the social hub after working hours. But with the advent of laptops and the tendency for people to consume entertainment separately, this has changed. 'The *Investigator*', says Don, 'will have three significant rec rooms with big TVs and couches, but I'll be very interested to see how much they are used, and what the social implications of that might be.'

It is, however, says Don, the importance of safety that has evolved most significantly over the past decade. With today's culture of risk assessments and inductions, so has the inevitable paperwork. These days every new person who sails undergoes a multi-stage introduction to the proper usage of safety gear, alarms, exit routes, life rafts etc. 'Before every task there's a "toolbox meeting" where we all discuss the risks inherent in the job, how they can be mitigated, whose role it is to do what, and where everyone has to be. It's certainly changed since I went across the Bight', he adds with a quiet laugh.

So, I enquire, what was it like in the old days? He pauses slightly, perhaps a little embarrassed. 'Well, for example', he says quietly. '*Southern Surveyor* had a fixed gantry at the back where the A-frame is now situated. There was a ladder up to it – no railing, no back hoops or anything – you'd climb up this ladder and go out along a narrow little catwalk, which only had railings to knee-height, then along an I-beam that went out over the stern. We'd climb it, sit in a little seat and lean forward to handle blocks to change the ropes – sometimes in the dark, 8 m high and 4 m off the end of the ship!' He shakes his head a little at the memory. 'You just wouldn't be able to do it nowadays.' Just as well, we agree.

Even so, he says, you can never guarantee everybody's safety all the time. Don still feels pity for one scientist from the University of New England who came aboard *Surveyor* to work on fish parasites, but succumbed to seasickness early in the voyage and never really recovered. Staggering, in his misery, through a 70 kg fire door, he jammed his thumb badly when closing it. His swollen, blackened thumb was quickly bandaged, but the next day 'he came through that same door the other way and did the same thing to his other thumb!' I don't know whether to laugh or wince. Don does both.

THE FUTURE

Toni Moate

Executive Director, Future Research Vessel Project

> *'Whatever happens onboard her, whether it be success or failure, I will actually always feel personally responsible.'*

Very few people can talk about ships with the enthusiasm of Toni Moate, so one might be excused for thinking the Executive Director of the Future Research Vessel Project has spent a lot of time at sea, getting her feet wet exploring the currents and the creatures of the oceans of our region and beyond. Not a bit of it. 'Professionally, I'm an accountant', she tells me, on a glorious Hobart afternoon looking across the sparkling Derwent River.

Toni went to sea on *Southern Surveyor* only once, on a twenty-four-hour trip, and if she thought her standing within the organisation would pull any rank at sea she was in for a shock. 'I was duped', she says. 'When I came onboard, I'd hoped to take one of the main berths, but I was told I was just a Johnny-come-lately and I was given a cabin with someone who had been to sea a lot more than me, and who'd already grabbed the bottom bunk.' As Toni explains, the motion of the ship can make negotiating a top bunk a difficult proposition, particularly at night, requiring considerable gymnastic ability. 'It was hard enough getting in, but in the morning I could hardly get out', she says.

The best way, she was later told, was to stand on a nearby desk, support your weight with your arms and swing your legs across into the bed. Not that this information would have been much use to her anyway. 'I'm not that tall', she says, with a hint of indignation. 'It took me ages to get out of that bed. I'm glad they didn't have cameras – as it was very unladylike.' Toni's trip south was mercifully brief, and one of the calmest in memory.

More than just about anyone, Toni can tell the story of the conception and development of the *Investigator*, the ship which will take the traditions forged by *Southern Surveyor* and the vessels that preceded her into the twenty-first century. What does she see as *Investigator*'s most significant advantages over her predecessor? 'For a start, she will go further', Toni begins, 'and that means there'll be bits of the ocean we can get to for the first time, particularly into the Southern Ocean – our backyard.'

Although not an ice-breaker, *Investigator* is rated ice-capable, meaning she will be able to venture as far as the Antarctic ice edge. 'It also has much more geoscience capability', Toni continues. She turns and glances towards the new ship, pointing out the brilliant white dome atop her mainmast, inside which *Investigator*'s almost two-ton radar apparatus will fire off and receive 800 pulses per second, reaching deep into the atmosphere, distinguishing sleet from snow from rain and gathering highly detailed 24/7 weather data. '*Southern Surveyor* had nothing like that', she tells me. *Investigator* will be equipped with far more atmospheric analysis capability. 'The *Surveyor* had a few sensors for looking at air quality, but now for the first time we have a dedicated aerosol laboratory.'

Merely listing *Investigator*'s capabilities would probably take most of the afternoon. Some of what Toni didn't include was her capacity to accommodate forty scientists as opposed to *Surveyor*'s thirteen, her new ability to collect weather data in remote locations of the Southern and Indian oceans, thus improving our understanding of cyclones and other weather events, and her comprehensive suite of marine geoscience equipment to improve our search capabilities for oil and gas reserves. She will also be better suited to take on international collaborations, such as searching for the shipwrecks of the World War II Battle of the Coral Sea with the scientists who discovered the wreck of the *Titanic*.

Before we get to the genesis of *Investigator*'s very ongoing story, however, I want to learn a little of Toni herself. As I prompt her to outline her career path thus far, I begin to feel like I'm watching a great engine slowly come to life within the form of this extremely focused individual. 'I started in 1991 as a project officer in finance', she says matter-of-factly before proceeding to list her subsequent achievements, a process which seems to go on for about half an hour, starting with 'a stint in personnel … coordinating support across the labs … executive officer coordinating reviews of all CSIRO's science activities … research project management … working with grants … resources' etc.

Somewhere along the way, Toni became finance manager for ten years, then assistant chief in charge of business development, communications, facilities, health and safety, resources, then into Marine National Facility management. Just when I think she's done, she throws in, 'Oh, I'm also a CPA and a graduate of the Institute of Company Directors' as a sort of afterthought. Not bad for a girl who left school at sixteen.

As one of those whose career has helped forge the bigger picture of Australian marine science, I ask Toni how she would sum up *Southern Surveyor*'s contribution. She answers unwaveringly. 'Up until *Southern Surveyor*', she says, 'we'd had predominantly oceanographic-based vessels. *Southern Surveyor* changed that. She expanded the concept of how the MNF has been able to deliver multi-disciplinary blue-water science for Australia, and now that's been built on even further with the *Investigator*.'

One aspect of this evolution has been, Toni believes, *Surveyor*'s role in bringing the various communities of marine science disciplines together for their mutual benefit, providing them with a common platform and a common interest. 'They started to work in quite a different way', she says. 'I think having *Southern Surveyor* built an *esprit de corps* across those communities, and I don't think the MNF would have had quite the same journey if we'd gone from the *Franklin* straight to the *Investigator*.'

This could sometimes be as simple as an appreciation of the uses of equipment, such as the gravity meter. 'It was worth about half a million dollars', says Toni, 'and the geoscientists thought they were the only ones with a use for it. Then when we got the various heads of the disciplines together, we realised it could provide a critical piece of data to the oceanographic community by its ability to measure the exact height of the ocean via gravity calculations.' This sort of rapport, Toni believes, allowed the various communities to evolve from competitors to collaborators.

I ask Toni when she thinks the very first conversation concerning the new vessel, tied up across the water from us, took place. She considers carefully. 'I'd say about 2005', and glances over her shoulder at *Investigator*'s gleaming white hull. It was then that *Southern Surveyor*, long-serving converted fisheries trawler, was reaching her mid-thirties and, like a family of siblings awkwardly contemplating their ageing parents, MNF would sooner or later need to address the question of *Surveyor*'s future.

Discussions began on what sort of ship would be needed to replace her, what she would be required to do and what she would look like. Approaches to government were made, and the ever-vexed subject of money raised. 'We got knocked back three times', says Toni, with a look that tells me she was deterred by this not one little bit and would have been quite prepared to be similarly disappointed four, fifty or a hundred times.

In May 2009, the government announced that it approved a new MNF vessel, and a commissioning and construction management team began to be assembled. The initial specifications were put together – 'a kind of wish list' – and expressions of interest put to the market. 'It was part of the nation-building funds of that year', says Toni, 'so we had to get it out pretty quickly.'

Toni balanced relationships between government, stakeholders and the vital MNF Steering Committee on which she sits. After a long tender process, detailed negotiations with a smaller group were commenced, and a construction contract was signed in January 2011.

Without a research ship construction bid from Australia, the build took place in Singapore. 'To tell the truth', says Toni, 'no one is actually a big research ship-builder – there just aren't that many research ships. And they usually last about twenty-five years or more, so, globally, there are only a handful being built at any one time, anywhere.'

With the contract signed, the practical considerations of what would in many ways be a concept design began to be digested and reduced to the fundamental notion of

'This is what we need the ship to do, show us what it would look like.' Over the life of the project, Toni tells me, there were more than 3000 separate drawings – a massive amount of information in which every part of the ship was provided with a detailed design which would need to be assessed, commented upon and altered. Likewise the many subsequent drafts, until the final construction version was arrived at.

At this stage, I'm prompted to tell Toni of a conversation I'd had recently with Sarah Schofield, CSIRO's communications advisor and the woman on whose shoulders rests the task of presenting the *Investigator* to the public. Sarah had recently given me an insight into the exquisitely simple logic occasionally involved in designing something as incomprehensibly complex as a modern ship. At a design meeting to discuss the media requirements of the new vessel, she outlined a vision of the ship being capable of sending images and stories of *Investigator*'s work live via satellite to a media-hungry world. As she spoke, she became of aware of notes being taken and looks being exchanged across tables, accompanied by some muttered nods of agreement before a quietly uttered comment, 'Yes, we can do that.' 'At that moment', Sarah told me, 'it struck me that I had just been responsible for enabling the scientists onboard the ship to communicate live to the world. It was quite daunting.'

'Yes, that's the way it sometimes happened', says Toni. 'Sarah might organise the live-via-satellite capability, then someone else might suggest another feature, but then all those components would have to be made to work together. That's been a huge part of it.'

In January 2013, the final block was lifted into place and the fit-out of RV *Investigator* began in earnest. The ship is 93.9 m long, has ten internal storeys and can sleep forty scientists and support staff, and twenty crew. Source: CSIRO/Chris Dickinson.

Toni Moate and the owner's representative, Graham Stacey, look on as the plasma cutter at the shipyard slices through a sheet of steel. Source: CSIRO/Ben Rae.

I had been shown some remarkable footage of the *Investigator* being constructed in Singapore, coming together on vast wheeled construction gurneys, one gargantuan prefabricated chunk at a time. I ask Toni if she made the trip to see it for herself. 'I ... probably went up there more than twenty times', she says, graciously not making me feel even the slightest bit foolish. A team of eight people from CSIRO oversaw the construction, she explains, working through the myriad technical issues over the life of the build.

The sheer scale of the logistics sounds utterly daunting. 'When we first cut the steel, that was a big event', she says. She explains how all the sheets were laid out in a shed, 'about two soccer field in size', before being marked with patterns on a computer with every last piece being utilised. 'Have you ever done dress-making?' she asks, a question I'm reasonably certain has never been directed at me before. 'It was just like a pattern. The same way you put a collar there and a sleeve there. Then, of course, the cut pieces have to be put somewhere where they can be found again when they need them.' At this point we both pause, barely attempting to contemplate the scale of the undertaking and grateful that there are people in the world who know exactly how to do such things!

The interest in *Investigator* from international science bodies is a factor of which CSIRO is acutely aware, even to the extent of sharing a few 'trade secrets'. When presenting to the International Research Ship Operators (IRSO) after signing the contract, Toni says they were 'blown away' by just how much ship was obtained for the price. As has been pointed out, Australia, despite its vast coastline, has traditionally had only a single dedicated blue-water marine research vessel, compared to countries like Canada which runs a whopping thirty-five, and even Belgium, with what Toni describes as by comparison 'a ridiculously small coastline', has six research ships to its name.

CSIRO has been approached by other countries to assist with their submissions for research ships, and Toni is happy to explain to her international colleagues the methods employed in making *Investigator* the diverse project she has been designed to be.

Toni's relationship with *Investigator* will always be a personal one, forged right at the beginning. 'I actually pushed the button to start the steel-cutting machine', she tells me with pride, then leans forward conspiratorially to let me know that she is *Investigator*'s 'lady sponsor', the female figure which in maritime tradition is attached to every new ship before undergoing sea trials. 'There's a naming ceremony and a blessing, and even a bottle of champagne across the bow.' 'Were you afforded that particular privilege?' I enquire. 'I was', she says, with a quiet glow.

As we finish, and the afternoon sun setting over Mount Wellington illuminates the top part of *Investigator*'s superstructure and her bulbous domed radar, Toni feels the need to emphasise once again just what the project, and now the ship, has meant to her personally. Looking at me solemnly, she says, as if uttering some kind of oath, 'I will feel probably forever responsible for the ship. I would never have thought this when I was given the project and it would be interesting to know if people who take on buildings and things like that feel the same way, but whatever happens onboard her, whether it be success or failure, I will actually always feel personally responsible.'

***

If there is one regret that I have in telling some of the story of the RV *Southern Surveyor* and the people who worked with her, it is in having properly seen her only once, in Hobart, when Mike Jackson led me down her empty decks and corridors after her final voyage as Australia's Marine National Facility. Having heard the stories of her many endeavours and discoveries, it simply didn't seem right that these same passageways – which it seemed I had almost come to know – should now be silent. Instead of my own echoing footsteps, I longed to feel the reverberations of her engines, the fearsome swell of the Southern Ocean or, gliding over a glassy Arafura Sea, the cloying heat of the tropics.

It is perhaps a little glib to speak of a ship – any ship – as one would a person, of her steel plates being 'special', 'unique' or possessing 'character'. A ship is, after all, little more than the sum of its people and its story. Let it be said of *Southern Surveyor*, however, that she elicited affection, that she gave purpose and meaning to the lives of many, that what was learned onboard her contributed – and will continue to contribute – to our nascent understanding of the oceans. The importance of this, surely, in the twenty-first century, can hardly be over-emphasised.

As much as possible, RV *Investigator* must be all things to all Australian marine scientists, as Australia has only one blue-water research vessel. The vessel will support atmospheric, oceanographic, biological and geoscience research. Source: CSIRO.

Unprompted, virtually everyone I met who spoke of *Southern Surveyor* did so warmly, and felt safe and productive within her confines. And yes, some did describe her as 'special'.

The scientists, technicians and crew I spoke to were without exception generous and forthcoming, all eager to tell of the part they played in *Surveyor*'s story. Many were obviously unused to speaking about themselves and their work, and after a little coaxing seemed touchingly honoured to be part of the project. In reality, it was I who was humbled by their passion, their knowledge and their quiet dedication to the truth of science.

List of interviewees

Richard Arculus

Richard Arculus' academic career has focused on the sources and evolution of magmas associated with island arcs and their adjacent backarc basins. He has explored many subaerial volcanoes but over the past 30 years has concentrated on submarine volcanic activity, particularly in the western Pacific. He has led many scientific expeditions to this region onboard Marine National Facility vessels.

John Barr

John Barr served as First Mate of the *Southern Surveyor* from 2008 to 2014 and as Master from 2012 to 2014, including the vessel's final voyage south of Tasmania. He has been at sea for 30 years and rates his time on the *Surveyor* as the most enjoyable and rewarding of his career. He is currently working in the oil and gas sector and lives in Dunedin, New Zealand, with his family.

Robin Beaman

Robin Beaman is a marine geologist based at James Cook University in Cairns, Australia. His research goal is to understand the relationships between the physical environment and the distribution of marine life on the seafloor, including the geological and oceanographic processes that have influenced the shape of Australia's underwater landscape.

Nathan Bindoff

Nathan Bindoff is a physical oceanographer, specialising in ocean climate and the Earth's climate system. His professional responsibilities include being Professor of Physical Oceanography at the University of Tasmania and CSIRO Marine Research Laboratories, Climate Change and Ocean Processes program leader, and Chief Investigator in the ARC Centre of Excellence in Climate System Science.

John Boyes ('Boysey')

John Boyes was involved in the New Zealand commercial fishing industry from 1965 to 1980, operating both his own and several company vessels. He later taught at the Australian Maritime College in Launceston, Tasmania, as Master of the fisheries training vessel, before becoming involved with the CSIRO research vessels *Soela*, *Franklin* and finally *Southern Surveyor* as First Mate. He is now retired from the sea.

Jeff Cordell

Jeff Cordell has been a senior technical officer in CSIRO for nearly thirty years. He was involved in the initial fit-out of *Southern Surveyor* when it arrived in Tasmania in 1989. Over the next few years he was heavily involved in the technical development of the science facilities and assisted in the development several new instruments for the vessel.

Patrick De Deckker

Patrick De Deckker FAA, AM is Emeritus Professor in the Research School of Earth Sciences at the Australian National University. One of his research interests is the past history of the oceans around Australia, which he has explored by taking numerous sediment cores using various vessels, including *Southern Surveyor*. While at sea he also trained students, future Australian marine geoscientists.

Martina Doblin

Martina Doblin is an Associate Professor at the University of Technology, Sydney. Having completed a PhD on growth of harmful algae in estuaries, Martina's more recent research has been in Australian coastal waters where she is investigating the ecological functions and ecosystem services of phytoplankton, the microscopic algae that form the base of the marine food web.

Seamus Elder

Seamus Elder settled in Hobart in 1968, where he worked for the Hobart Marine Board. He later joined the Navigation and Survey Authority of Tasmania as Surveyor, and Examiner of Engineers. In 2002 he moved to the Australian Maritime College as engineer on its training ships. He was an engineer on the *Southern Surveyor* from 2004 until she was decommissioned.

Neville Exon

Neville Exon has been a geoscientist since 1963 and has produced nearly 200 geoscience publications. Most of his time has been spent at the Australian Geological Survey (now Geoscience Australia) and he is now at the Australian National University, where he is responsible for Australian and New Zealand involvement in scientific ocean drilling. He has participated in about fifty marine geoscience surveys.

Brian Griffiths

Brian Griffiths joined CSIRO in 1970 as curator of the zooplankton collection. He worked on the zoogeography of various zooplankton and on trophodynamics before moving to Hobart in 1985. From 1985 to retirement in 2014, Brian researched factors controlling primary production in the Southern and equatorial oceans, and was involved in the design of the *Investigator.*

Ian Hawkes

Ian Hawkes is currently the team leader for the CSIRO Oceans and Atmosphere Data Acquisition and Processing team, which is responsible for managing the information and communication technology systems as well as the scientific data logging systems on the Marine National Facility research vessel. Ian graduated with a BSc (Hons) in Physics from the University of Tasmania in 1986 and previously worked in the defence and aerospace industries.

Andrew Heap

Andrew Heap is a group leader at Geoscience Australia and has over fifteen years' experience leading marine geoscience research. He has published over 100 scientific and technical papers, and his research established seabed geomorphology as a criterion for design of Australia's current Commonwealth marine reserve network.

Tom Hubble

Tom Hubble has been investigating the geology of marine, estuarine and riverine environments since 1984. He is based at the University of Sydney and has published several papers on river-bank collapse and the stabilisation of river-banks against collapse by tree roots. He is currently investigating the causes and consequences of eastern Australia's submarine landslides.

Rudy Kloser

Rudy Kloser is a senior research scientist and leader of the acoustic and pelagic ecosystem team at CSIRO's Oceans and Atmosphere Flagship. Rudy investigates reflected and ambient acoustic signals in the deep ocean with associated optical and physical sampling to make ecological inferences about the water column and the seabed, and learn about ecosystem function and dynamics.

Kel Lewis

Kel Lewis first went to sea in 1981 as a greaser onboard a bulk carrier and since then has worked on all kinds of ships, from tankers to pipe layers and cargo. He started working on the *Southern Surveyor* in 2008 and it fast became his favourite working platform. Kel has accepted a role working onboard the new Marine National Facility research vessel *Investigator* as a member of the crew.

Mark Lewis

Mark Lewis has worked for CSIRO since 1990 in the area of fisheries research, mainly in stock assessment and habitat mapping, as a seagoing biological technician.

On the *Southern Surveyor*, Mark specialised in the design and use of sampling equipment, as well as sorting the biological samples obtained. His work has been recognised by two awards.

Don McKenzie

Don McKenzie joined CSIRO in 1977 as an agricultural field technician. In 1989 he moved to Hobart and for nine years worked as a marine technician on a range of projects and vessels – but mostly on *Southern Surveyor*. In 1998 he moved into the ship management team and is now science operations manager for the Marine National Facility. Don made around forty voyages on *Southern Surveyor*.

Tara Martin

Tara Martin is a marine geophysicist with a speciality in seismic activity, marine acoustics and tectonics. She has worked on research vessels around the world since 2000. She joined CSIRO in 2010 and now works as the Research Group Leader for Data Acquisition, Processing and Management.

Steven Micklethwaite

Steven Micklethwaite researches how the Earth deforms across multiple scales, and particularly how these processes influence mineral resources. He received his PhD in Geology from Leeds University, UK, in 2002. In 2010 he was the recipient of the Rising Stars Award, and in 2011 was the Hammond-Nisbet Fellow. In 2012 he was a co-scientist on the ECOSAT oceanographic survey of the Coral Sea.

Toni Moate

Spanning a career in CSIRO of more than 20 years, Toni Moate is Strategy Director, National Facilities, and Executive Director, Future Research Vessel (FRV) Project. Toni managed the Marine National Facility from 2009 to 2014 and, as Executive Director of the FRV Project, was responsible for replacing *Southern Surveyor* in the MNF with the newly constructed *Investigator*.

Tim Moltmann

Tim Moltmann is the Director of Australia's Integrated Marine Observing System, based at the University of Tasmania. He worked at the CSIRO for over a decade, rising to be Deputy Chief of the Marine and Atmospheric Research Division. He has been Australia's representative on international resources forums for the Indian Ocean and tropical Pacific observing systems.

Alicia Navidad

Alicia Navidad works as a senior hydrochemist for CSIRO. Her speciality is method development for high-accuracy nutrient analysis including QA/QC protocols, software functional specification and documentation. She managed the *Southern Surveyor*'s laboratory and has been the lead for Antarctic voyages in both logistics and high-level accuracy data output.

Lindsay Pender

Lindsay Pender started his career as a nuclear physicist working in Canada and at the Australian National University. He then transferred to CSIRO and in 2008 started working as part of the computing support team for the *Southern Surveyor*. Lindsay used his experience in oceanographic research, IT systems and engineering to help design the RV *Investigator*.

Ron Plaschke

Ron Plaschke joined CSIRO in 1984 as a seagoing analytical chemist and has since been involved in small boats, diving, ships operations and management. In 2009, Ron was appointed Director, Marine National Facility, and guided by an independent Steering Committee has overseen the transition from *Southern Surveyor* to the new research vessel *Investigator*.

Fred Rostron

Fred Rostron joined the *Southern Surveyor* as Chief Engineer from a varied background in the merchant navy. He was involved in the running and upkeep of the vessel and the supply of services to keep scientific equipment functioning.

Matt Sherlock

Matt Sherlock has worked at CSIRO for more than twenty-five years as an electronics engineer. He was a member of the project team which converted *Southern Surveyor* to a science research platform in the late 1980s. Matt enjoyed a close involvement with *Surveyor* throughout its life with CSIRO, developing subsea systems for scientists and participating in many memorable research voyages.

Bernadette Sloyan

Bernadette Sloyan is a Principal Research Scientist and Group Leader with CSIRO's Ocean and Atmosphere Flagship. She is the Co-Chair of the international Global Ocean Ship-based Hydrographic Investigations Program (GO_SHIP) and is an international leader in documenting and understanding the role of ocean circulation and variability in the global climate system.

Iain Suthers

Iain Suthers is a fisheries oceanographer at the University of New South Wales and is based at the Sydney Institute of Marine Science. He studies the 'physics to fisheries' ecosystems of estuaries, artificial reefs and the East Australian Current. He had five voyages on the RV *Franklin* between 1993 and 1999 and after 2004 he led six voyages on the RV *Southern Surveyor* into the Tasman Sea.

Ian Taylor

Ian Taylor joined the *Southern Surveyor* in 1991 and served as her Master. He witnessed the shift in focus from primarily fisheries research to multi-disciplinary research. Prior to that he worked on product tankers on the Australian coast and on bulk carriers internationally.

Steve Thomas

Steve Thomas is an electronics technician and manages the scientific electronics support team for the RV *Investigator*. He joined CSIRO in 2001, after working under contract with the Royal Australian Navy and Geoscience Australia. In the past at CSIRO, Steve conducted seagoing and vessel dry-docking electronics support, with the RV *Franklin* then *Southern Surveyor*.

Tom Trull

Tom Trull received his PhD from the MIT – Woods Hole Joint Program and joined the Antarctic Climate and Ecosystems CRC in Hobart in 1993. He leads projects examining oceanic control of atmospheric carbon dioxide levels, including establishing the Southern Ocean Time Series of automated climate and carbon cycle observations.

Anya Waite

Anya Waite is a biological oceanographer whose primary research interests are the links between ocean physics, biology and biogeochemistry. Following terms at the Woods Hole Oceanographic Institution and the University of Western Australia, Anya is now a professor at the Alfred Wegener Institute in Germany.

Chris Wilcox

Chris Wilcox is a research scientist with CSIRO Oceans and Atmosphere Flagship in Hobart, Tasmania. He has a Masters degree and a PhD in ecology and conservation biology. He has worked with NGOs, government and private enterprise over a twenty-three year career as a professional biologist to develop cost-effective solutions to natural resource management problems.

Colin Woodroffe

Colin Woodroffe is a geomorphologist in the School of Earth and Environmental Sciences, University of Wollongong, with a PhD and ScD from the University of

Cambridge. He was a lead author on the coastal chapter in the 2007 IPCC Fourth Assessment report, and has written several books on coasts.

Lisa Woodward

Lisa Woodward joined the CSIRO Marine National Facility in 2007 as an Administration Support Officer. She later moved into Voyage Operations and voyage manager roles on the *Southern Surveyor.*

www.ingramcontent.com/pod-product-compliance
Lightning Source LLC
LaVergne TN
LVHW052344100826

845147LV00012B/751

* 9 7 8 1 4 8 6 3 0 2 6 4 2 *